TOBIAS MAYER'S *OPERA INEDITA*

Tobias Mayer

[*Frontispiece*

Plate I

Tobias Mayer's *Opera Inedita*

THE FIRST TRANSLATION OF THE LICHTENBERG EDITION OF 1775

ERIC G. FORBES
University of Edinburgh

American Elsevier Publishing Company, Inc.
New York 1971

Published in the United States by
AMERICAN ELSEVIER PUBLISHING COMPANY, INC.
52 Vanderbilt Avenue, New York, New York 10017

International Standard Book Number: 0-444-19578-5

Library of Congress Catalog Card Number: 79-173863

Printed in Great Britain

CONTENTS

	page
Preface	vii
Introduction	1
The life and work of Tobias Mayer (1723–1762)	1
Georg Christoph Lichtenberg (1742–1799) and the editing of Tobias Mayer's unpublished works	15
Translator's notes on Mayer's treatises	21
Opera Inedita	
Preface	51
Treatise I. A More Accurate Definition of the Variations of a Thermometer	53
Treatise II. Astronomical Observations Taken on the Mural Quadrant in the Observatory at Göttingen	62
Treatise III. A Simple and Accurate Method of Computing Solar Eclipses Visible in a Given Place	71
Treatise IV. Treatise on the Relationship of Colours	81
Treatise V. Treatise on the New Catalogue of Zodiacal Stars	92
Treatise VI. Treatise on the Proper Motion of the Stars	109
Commentary on I	113
Commentary on II	120
Commentary on IV	122
Remarks on the Selenographic Table	136
Bibliographical References	141
Name Index	163
Subject Index	165

PREFACE

The researches into the theory of the Moon's motion carried out around the middle of the eighteenth century by Alexis Clairaut, Leonhard Euler, and Jean D'Alembert have been recognized by historians of astronomy as laying the foundations of a very fertile phase in the development of celestial mechanics which owed its success to the fruitful application of the new analytic methods to the geometrical principles of Newtonian mechanics. By the end of that century, it appeared that Newton's axioms of motion and his hypothesis of an inverse-square law of gravitational attraction were capable of providing a sufficient basis for all the planetary motions and perturbations, and for the numerous observed irregularities in the Moon's motion; while the brilliant theoretical contributions of Joseph Louis de Lagrange and Pierre Simon de Laplace gave rise to a general confidence in the long-term stability of the solar system without the need to introduce the transcendental hypothesis of a Divine Being—so essential a feature of Newton's own providential explanation of the structure of the solar system. This optimism, of course, turned out to be unjustified, for subsequent advances in the mathematical structure of the lunar theory led Denis Poisson to discover that Laplace had *not* succeeded in accounting quantitatively for the phenomenon of the secular acceleration of the Moon's mean motion. The effects of tidal friction and other geophysical and meteorological factors are now known to produce small but appreciable irregularities in the Moon's motion and in the rate of the Earth's own axial rotation, which impose an intrinsic limitation on the power of analysis and consequently in the predictive ability of the lunar theory itself.

At a time when discrepancies between the lunar theory and observation were being generally attributed to the approximate nature of the law of gravitation or to the existence of a hypothetical resisting aetherial medium, Tobias Mayer was insistent that the cause lay rather in the weakness of the observational data themselves. It was this attitude—so typical of a practical astronomer—which led him in 1753 to recognize the fundamentally important fact that the principal cause of the discrepancies of 5′ and more in the respective theories of Clairaut, Euler, and D'Alembert was the reflection, in the observed celestial co-ordinates of the Moon, of errors in the observed star positions. This conceptual breakthrough arose from his combination of two independent methods of finding longitude differences, with which he had been closely concerned throughout the previous seven years; namely, observations of lunar occultations and of lunar and solar eclipses.

By comparing the predictions of the celestial positions of the Moon's conjunctions and oppositions with the Sun, based on his observations of lunar occultations, with the corresponding positions derived from data which he collected for all the lunar and solar eclipses observed since the invention of the astronomical telescope and pendulum clock, Mayer corrected the value of the Moon's mean motion determined from these occulations and thereby calculated the errors which had been inherent in the star positions. Having eliminated these in the case of over 100 zodiacal stars, whose celestial co-ordinates could now be relied upon to between 5″ and 10″, he used those star positions to re-calculate the Moon's mean motion. Simultaneously, he improved the internal accuracy of his theory by taking account of a larger number of terms in the infinite series which it contained. The result of these improvements was that he now felt able to predict the Moon's longitude to within 30″, and recognized that any further increase in accuracy would have to come from refinements in practical techniques, better instrumentation, and in the assimilation of new and more reliable observational data. Thus it is easy to understand why Mayer's subsequent astronomical researches were devoted to:

(i) the elimination of instrumental errors in a mural quadrant;
(ii) a simple and accurate method of computing solar eclipses;
(iii) the construction of a catalogue of zodiacal stars; and
(iv) an investigation of the proper motion of stars.

These are, in fact, the respective topics of four of the six treatises contained in this present work, the only volume of Mayer's unpublished writings to have been printed prior to this time. A fifth treatise, composed in 1755 and devoted to the problem of analysing the fluctuations of thermometer readings (so necessary for the quantitative determination of the effect of astronomical refraction), would appear to owe its inspiration to the success which its author had had in accounting for the numerous periodic inequalities in the theory of the Moon's motion, and which he was also having in similar investigations into the motion of the planet Mars. The remaining treatise is concerned with a mathematical theory of colour-mixing. His researches in this field could be regarded simply as a hobby which Mayer took up when the French occupation of Göttingen during the Seven Years' War curtailed his activities as an observer; but they might equally well have been due to his concern with the wavering fortunes of the Cosmographical Institute for which he had formerly worked in Nuremberg and which had now been re-established in Göttingen. Mayer's precepts and empirical rules for preparing reproductions of heraldic symbols and maps, were capable—at any rate in principle—of being followed by the untrained apprentices who were having to be employed in lieu of the many skilled artists and cartographers who had been unable to take part in the transfer. Lichtenberg himself would appear to have been ignorant of the above-mentioned inner connections among the six treatises which he edited, and of their mutual link with Mayer's preceding studies in the lunar theory. Certainly he made no explicit reference to them as criteria for the particular selection published. The simplicity and hence the speed of editing appear to have been the primary factors governing his choice; secondary factors being that the six Latin treatises (or lectures) in question had all been read by Mayer to the Göttingen Scientific Society at regular annual intervals throughout the period 1755 to 1761 inclusive, and are all comparable in length.

Eric G. Forbes

30 January 1970

INTRODUCTION

The Life and Work of Tobias Mayer (1723–1762)

(Johann) Tobias Mayer (Fig. 1, Plate 1), whom posterity remembers for his researches in astronomy and in particular for those concerned with the theory of the Moon's motion, was first and foremost an artist with a fascination for military affairs. He was born in Marbach, a small town on the river Neckar only a few miles from Stuttgart, on 17 February 1723 but his family moved to the neighbouring town of Esslingen while he was still an infant,[1] and it was there that he spent the rest of his childhood and youth. He was only eight years old when his father died,[2] leaving his mother and her six surviving children[3] in a desolate condition. It would appear that Tobias, being the youngest, was then sent to the local orphanage where he was to spend the remainder of his childhood years.[4] His mother found work as a hospital orderly at the St Katherine's Hospital in Esslingen—a circumstance which gave Tobias the opportunity to acquire a knowledge of the architectural lay-out of that building. A plan-drawing of it, which he prepared before he was 14 years old,[5] greatly impressed an elderly non-commissioned officer in the Swabian District Artillery then stationed in Esslingen—a fusilier by the name of Geiger—who immediately sought and obtained the town council's permission to instruct the boy privately in the arts and crafts of war.[6] Two important results of this tuition were that:

(a) it roused a latent ambition in Tobias to become a military engineer which, fortunately for science, was destined never to be fulfilled, although he taught the various aspects of artillery to other boys in Esslingen and later lectured on this subject to students at the University of Göttingen;[7]
(b) it introduced him to mathematics, and in particular to projective geometry, which did not feature on his school curriculum.[8]

Historical data point to the fact that he quickly recognized the great value of analytical and projective geometry, especially in helping to solve many of the problems in ballistics and in military fortification; and he avidly read as many mathematical books as he could (often until the early hours of the morning). The rector of the Latin school which he was then attending encouraged Tobias in this pursuit of knowledge by allowing him access to his private library, and the youth found in a local shoemaker—a highly intelligent man named Gottlieb David Kandler, whom he always held in great respect—someone with whom he could share his new-found enthusiasm.[9] Mayer's first published work, *A New and General Way of Easily Solving all Geometrical Problems by Means of Geometrical Lines*

(Esslingen, 1741)[10] completed on the occasion of his eighteenth birthday, is essentially a demonstration of the application of analytical methods to the solution of common geometrical problems. Significantly, it ends with the problem of inscribing an irregular polygon in a circle, which was then of great practical importance in the design of military fortifications; the last of ten problems which are left for the reader to solve by the method described in the third and final section of his book, concerns the fortification of a heptagon whose sides have to be proportioned in accordance with certain prescribed algebraic conditions.[11]

On coming of age in 1744, Mayer left Esslingen and took up employment in the copper-engraving and map-making firm of Johann Andreas Pfeffel in Augsburg. It was during his two-year stay here that he became well acquainted with the optician and mechanic Georg Friedrich Brander, with whom he collaborated on the construction of the glass micrometer which he employed with considerable skill and success in several of his later researches.[12] A splendid index of the extent to which he had by this time developed both his natural artistic talent and his scientific knowledge is provided by his impressive *Mathematischer Atlas* (Augsburg, 1745) in whose 60 plates he presents the whole range of elementary mathematics with (in turn) astronomical, geographical, military, architectural and mechanical applications. This is the first positive evidence of his knowledge of astronomy and geography, which he must have been acquiring during the course of his last three years in Esslingen. He illustrates the use of projective methods in representing lunar and solar eclipses, and occultations of planets and stars by the Moon; and twice (cf. *ibid.* Tab. *XXIV and XXV*) makes reference to the fact that he already possessed manuscript solar and lunar tables prepared by himself.

Mayer's interest in the Moon's topography and in its motion originated somewhat later, however, after he left Augsburg and went to work at the Homann Cartographic Bureau in Nuremberg: it was roused by his attempt in 1747 to make an orthographic projection of the lunar surface with the intention of discovering which markings would gradually be covered by the Earth's shadow. This was a preliminary to a graphical construction of a partial lunar eclipse in the night between 8 and 9 August 1748, which was to serve as the basis for the determination of terrestrial longitude differences. He soon realized that the best of the existing lunar maps—those of Hevel and Riccioli—were not reliable enough for this purpose, and consequently he resolved to prepare a new map himself. It was with this end in view that he began, early in 1748, to make a regular series of observations, with his glass micrometer, of a considerable number of large and small features on the Moon's disc.[13] The technique which he adopted was to align his micrometer in the focal plane so that its parallel rulings were themselves parallel to the direction of the Moon's apparent motion. Over a short interval of time, the change in the Moon's declination would be negligible and the perpendicular lines would therefore correspond to the apparent hour circles; in practice, however, Mayer reduced his measurements to the true hour circle to eliminate any small systematic error that would otherwise have arisen on this account. Having clamped his telescope firmly with the micrometer in this position, he recorded the mean solar time at which the leading limb of the Moon crossed one of the perpendicular lines, noting, whenever possible, the micrometric position of the north or south limb at the same instant. The timing was actually done by his having a simple pendulum with a lead weight in a position which was convenient for him to observe, the time of whose oscillations he compared immediately before and after his observations with a seconds-clock in the neighbouring room. The estimated times are quoted by him on some occasions to the nearest 0·1 second, which is indicative of the high precision he was attempting to attain. When the leading limb of the Moon was not visible, he timed the passage of the horns through the perpendicular line.

From these measurements he was able to calculate the Moon's diameter in cases where the line joining the horns was at a very acute angle to the micrometric rulings; otherwise, a little before or after making his observations, he set the micrometer so that the perpendicular line passed through the two horns of the Moon, and estimated its diameter from the distance between the parallel rulings. With this information, he could now determine the positions of the lunar markings uniquely on the Moon's disc in terms of the *absolute* angular distances from two adjacent limbs and in terms of the *relative* distance from the disc centre and the angle made at the centre with respect to the north direction. Thus the effect of the Moon's variable distance from the Earth could be eliminated. The elaborate procedure by which Mayer systematically rids his results of all further dependency upon the terrestrial observer is described by him in a lengthy treatise on the Moon's libration, published in Nuremberg in 1750,[14] which is noteworthy as being the first example of the use of selenographic co-ordinates to specify the Moon's topographical features. The selenographic latitude was easily found by referring the positions of the lunar markings to the Moon's equator, analogous to fixing terrestrial latitudes. Mayer selected as the prime meridian of selenographic longitude, that which lies in the same plane as the line joining the centre of the Moon to the position of the *mean* Earth. Since the Earth's variable motion around the Sun was known, the relation of the mean to the actual position could always be easily calculated.

An interesting feature of Mayer's analysis of the Moon's libration, which was based principally upon his observations of the small but well-defined crater Manilius, is that he correctly formed three normal equations from a system of 27 conditional equations—a procedure never before attempted, for which no theory had at that time been developed.[15] His success in distinguishing the respective contributions of the irregularity of the Moon's orbital motion, its variation in angular diameter, and the effects of parallax, precession, and nutation, encouraged him to embark upon the preparation of 15-inch diameter lunar globes.[16] For various reasons, however, this project was never completed.[17]

The selenographic co-ordinates which Mayer had deduced from his observations on the three craters Manilius, Dionysius and Censorinus, served as fixed reference points for his visual estimates of the positions of other lunar markings which he sketched on planispheres with such care that he spent several hours drawing an area only one-hundredth part of the whole lunar disc. Fig. 2 (Plate 2) is the first of 40 such drawings from which the complete lunar map, of which Fig. 3 (Plate 3) is a photolithographic reproduction, was constructed. Fig. 2 is of interest in showing the relation between the number of parallel rulings and their angular separations, and reveals that Mayer in two cases was even estimating the latter to the nearest 0·01″. The lunar map which appears at the end of this volume is actually a copper-engraving of Mayer's half-scale copy of that illustrated in Fig. 3;[18] the diameters of his original drawings being 20 cm and 40 cm respectively.

Despite the minor imperfections to which Georg Christoph Lichtenberg draws attention in his commentary, the respective features of the Moon's topography were nevertheless easily identifiable, and the two maps will always bear testimony to Mayer's rare ability to combine practical skill with theoretical understanding. This talent was clearly appreciated by Abraham Kästner when he remarks that Mayer

'shared with others the mastery of practical mathematics, and realized that this resulted in a benefit to himself, since in general he judges sciences according to the degree of their kinship with utility. Many, who profess to be practical mathematicians, would substitute practice for theory altogether, but Mayer perceived with deeper insight that practice, which is strengthened in one way or another by common theory, must be freed from faults and perfected. Hence he both comprehended the theory of however great mathematical art of investigation the greatest intellects of our age had abandoned, and transferred it happily to practice.'[19]

The scholarly reputation which Mayer won for himself both through his original astronomical researches and through his skill and industry as a cartographer,[20] was responsible for his being called in November 1750 to a professorship at the Georg-August Academy in Göttingen,[21] an appointment which he duly took up at Easter of the following year. His title, Professor of Economy, was only a nominal one, for in his letter of

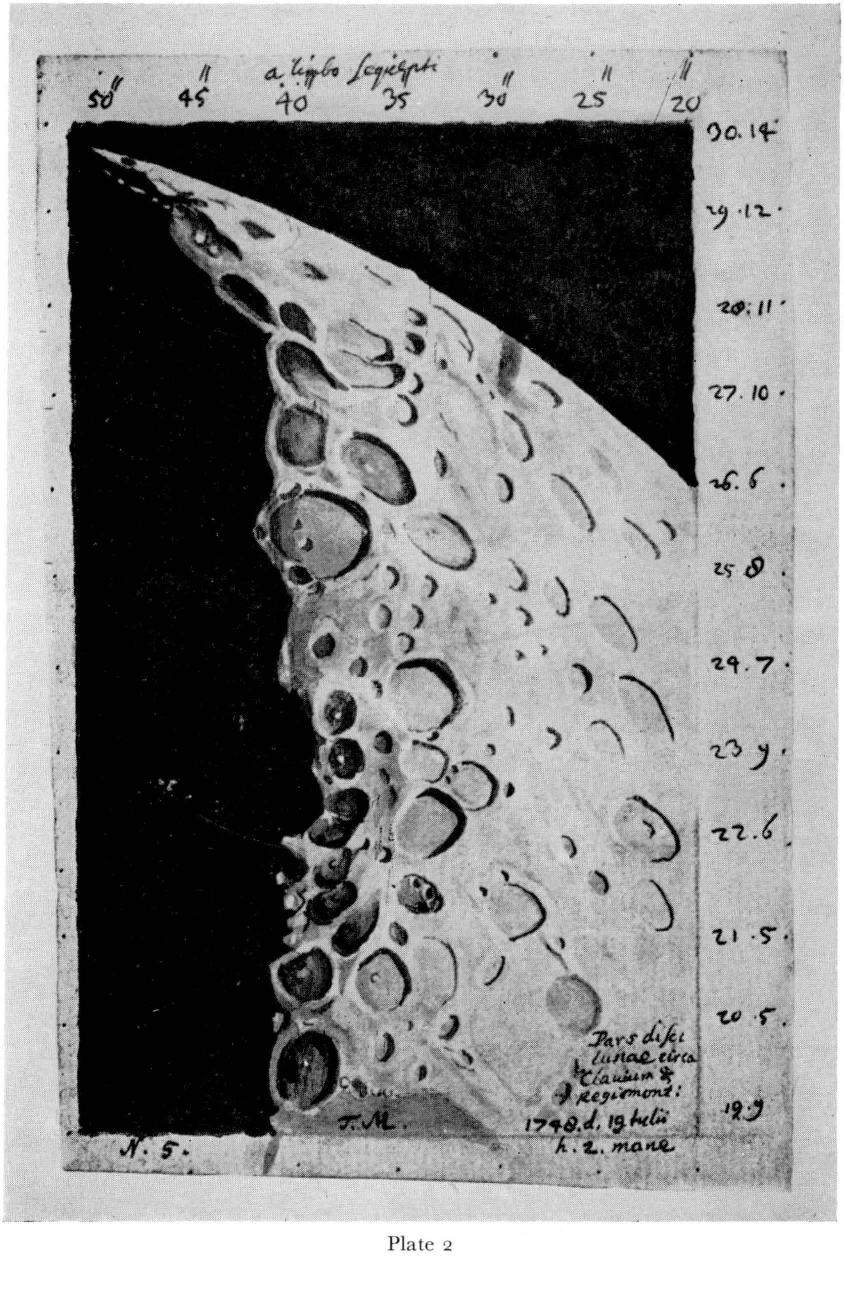
50
45
40
35
30
25
20
30.14
29.12
27.10
26.6
25.8
24.7
23.7
22.6
21.5
20.5
19.9
Pars disci
lunae circa
Clavium &
Regiomont:
1748.d. 19 Iulii
h. 2. mane
N. 5.
T. M.

Plate 2

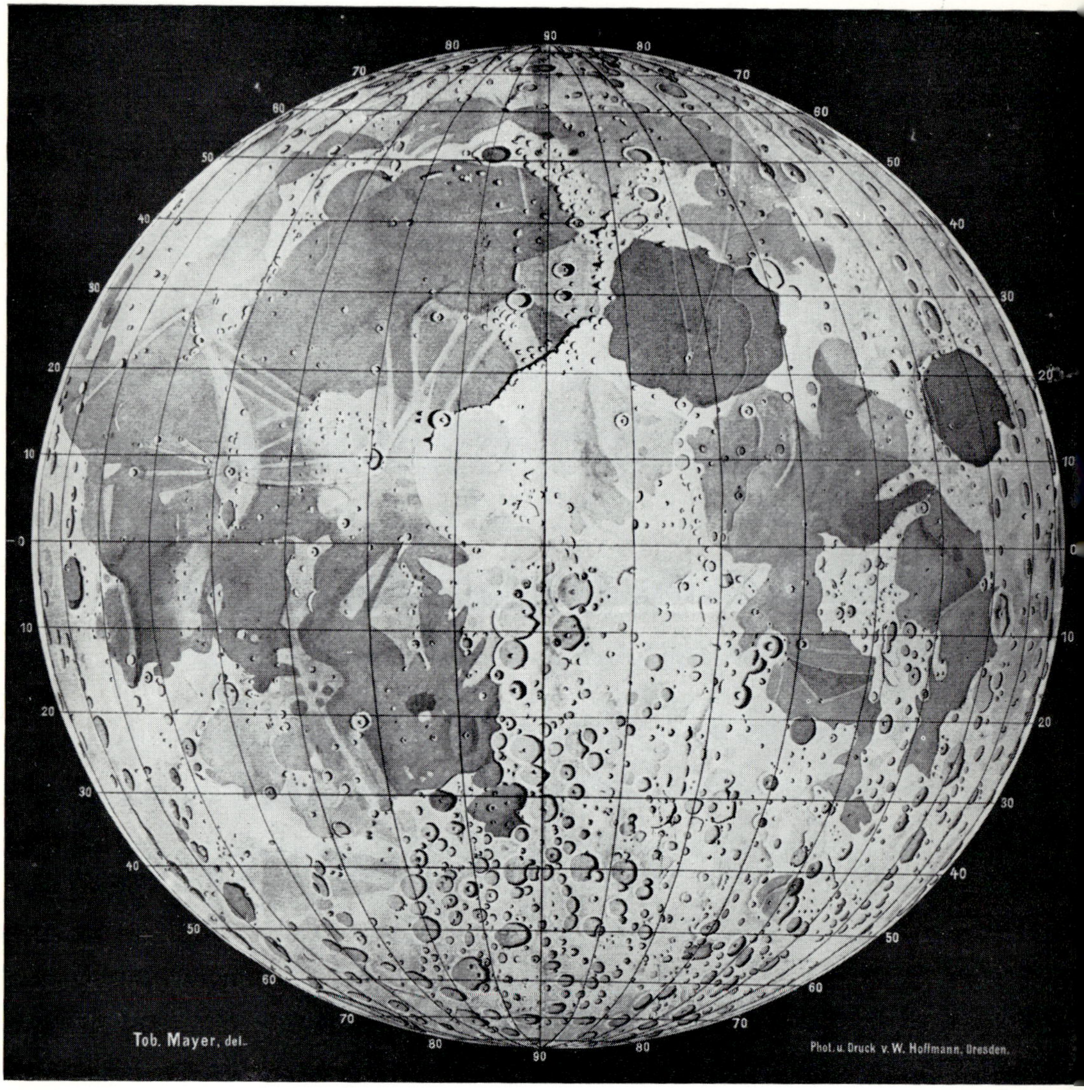

Plate 3. Photograph of a photolithographic reproduction of Tobias Mayer's larger lunar map (the original was 40 cm in diameter)

appointment his duties are stated as being the teaching of practical mathematics and research.[22] He brought with him to Göttingen his new bride Maria Victoria (née Gnüge), the daughter of a clergyman in Neckarbischofsheim, and the youngest sister of the wife of Johann Michael Franz, the Director of the Homann firm. She had probably been living with her relatives in Nuremberg after her father's death in 1747.[23] The marriage was apparently a happy one, and was certainly fruitful, five sons and three daughters being born to the Mayers in Göttingen within the nine-year period from May 1752 to May 1761;[24] of these, however, only four male children survived infancy and of the latter, only the two eldest reached the age of manhood.

After settling down in his new home, Mayer lost no time in initiating a correspondence with Leonhard Euler, for whose work he possessed a great admiration, and the 31 letters known to have passed between them during the course of the next four years constitute valuable source material on the development of lunar theory and the theory of astronomical refraction. Another theme which dominates the later phase of this correspondence is the possibility of Mayer's taking up a professorial appointment at either the Berlin or St Petersburg Academies, for both of which Euler was at that time acting as an intermediary.[25] Contrary to Mayer's expectation, his request to be allowed to resign his post in favour of the former invitation was not granted. Instead, he was asked to stipulate the conditions under which he would prefer to remain in Göttingen and—much to his surprise—was granted them. They included his receiving the sole directorship of the Göttingen Observatory, then in its final stages of construction, which he had been sharing since 1751 with Professor U. von Segner. By October 1754, Mayer had made up his mind to remain in Göttingen and confided to Euler that he had been thought of in Hanover more highly than he deserved.[26]

The main reason for Pierre von Maupertuis' invitation to Mayer to become a member of the Berlin Academy,[27] and for the determination of the Hanoverian Government—or more directly, of King George III (whose father had founded the Georg-August Academy in 1737)—to retain his services at all costs, was the international fame which Mayer had suddenly achieved as a result of his publication in volume II of the *Commentarii* . . . (or transactions) of the Göttingen Scientific Society of a set of lunar tables, for which he claimed an accuracy of a little more than 1′.[28] This constituted a significant improvement on Euler's own *Theoria motuum lunae* and the prize-winning treatise of Alexis Clairaut,[29] both of which had been published at the expense of the Berlin Academy in the course of the previous year. The following extract from Euler's letter to Mayer of 11 June 1754[30] is worth quoting since it not only reveals Euler's own view of Mayer's achievement but also reflects the state of informed opinion in Europe regarding it.

'Now, to come to your new important discoveries, I first of all congratulate you on them from my whole heart, and hope that everyone will soon be aware of their importance. Your first tables have indeed given rise to both approval and astonishment, only envy has again shown itself extremely clearly. You have undoubtedly seen M. d'Alembert's new treatise on the Moon, in which he refers to the tables in a disdainful manner and mocks the accurate agreement claimed for them: he does not believe that you were helped by the theory and supposes that he and M. Clairaut were the first to investigate the lunar theory . . . You would not be doing a bad thing if you were to take an opportunity of refuting this braggard in order to restore the reputation of the Germans. The reaction in England would appear to have been fairer, since such is the answer that I have received to the representations which I made about the importance of your tables, that I can conclude with [good] reason that you will be regarded as a worthy competitor for the prize offered for the discovery of longitude. If you could only now develop a method for determining at sea the position of the Moon through its distance from the fixed stars so accurately that they do not differ by more than half a degree from the truth when you compare them with the longitudes printed in your tables, you can be assured of the £20 000 prize. It would, however, also be necessary to determine more accurately the true positions of more fixed stars. Since, through your untiring labours, you have developed things to such an extent that your tables show the Moon's longitude to 30″ and the latitude to even 10″ accuracy, there is nothing better to be wished for from this side [i.e. from a theoretical standpoint], and M. d'Alembert himself admits that his tables are still in error by over 5′.'

The prize to which Euler is here referring is the largest of three bounties offered under the terms of the Act 12 Anne (1714) 'for Providing a Publick Reward for such Person or Persons as shall Discover the Longitude at Sea.' Mayer did, in fact, publish a paper in 1754 in the third volume of the Göttingen transactions containing an improved set of lunar tables and an explanation of the manner in which these could be applied to the solution of this problem—the basis of his method being the measurement of the angular distances between the Moon and a zodiacal star[31]—but he still remained sceptical of the accuracy which could be attained in practice for observations made on board a ship at sea. He was also of the opinion that the English would not be impartial towards claims made by foreigners. However, after having received this encouragement from Euler, he agreed to allow his colleague Johann David Michaelis, then Secretary for Hanoverian affairs in Göttingen, to sound out English opinion via the representative of the Hanoverian government in London, William Philip Best.[32] When the latter's reaction was found to be favourable, Mayer prepared a Memorial which was communicated along with related documents to Lord Anson, the President of the Board of Longitude which was directly concerned with the administration of the parliamentary rewards.[33] The papers were referred officially to Dr James Bradley, Astronomer Royal, who strongly recommended them (on two occasions) as a suitable basis for longitude determination at sea.[34] A navigational instrument—the Repeating Circle—which Mayer had designed for the purpose of making the necessary astronomical observations was, however, rejected since it appeared to offer no appreciable advantage over the Hadley octant.[35] No immediate decision was taken on the merits of Mayer's claim, largely due to the fact that the Board of Longitude was by that time committed to encouraging the alternative method of longitude determination which John Harrison's marine timekeepers promised to make practicable, and to Britain's subsequent involvement in the Seven Years' War (1756–1763) which prevented Mayer's method being tested at sea.

The completion of the Göttingen Observatory in 1754, and the installation there in the autumn of 1755 of an excellent six-feet radius mural quadrant that Mayer had had made by the highly skilled English instrument-maker John Bird,[36] placed him in a better position to make important contributions to positional astronomy, although his previous observaions in Göttingen which provided the basis of his lunar theory, were by no means insignificant.[37] With the aid of this new quadrant and a pendulum clock, he made a large number of observations of the zodiacal stars from the years 1756 to 1761; and he greatly improved practical techniques in astronomy by carefully investigating and quantifying the effects of observational and instrumental errors. His correction formula for meridian transits of stars, which came to be known by his name, was still being regarded as the most useful in existence as late as the end of the nineteenth century. He was the first person to recognize the necessity of correcting the meridian zenith distances—hence the declinations—of stars for the effect of the warping of a quadrant such as his own which, like most other instruments of this nature, was designed so that the axis of the viewing telescope lay appreciably outside the plane of the instrument.[38] Perhaps it was this discovery which gave him the idea of attaching a spirit-level to the viewing telescope that he introduced as one of his improvements to the astrolabe used in surveying;[39] for this ensured that the planes of the viewing telescope and of the graduated circular limb would each be parallel to the one plane (viz. that of the horizon).

The year 1757 marks a significant decline in Mayer's observational work in astronomy since, as a result of the French occupation of Göttingen from July 1757 until after his death in 1762, he was obliged to devote himself to studies which did not involve his leaving home during the night. The most time-consuming of these were his researches on the mixing of colour pigments[40] and on terrestrial magnetism,[41] although he was also continuing to give courses of lectures on various aspects of architecture, artillery, astronomy, cartography, mathematics, mechanics and optics.[42] Several reasons may be suggested for his interest in developing a mathematical theory of colour-mixing: it may have been simply a further expression of his natural artistic talent; or a response to the demand for a greater degree of consistency in the reproduction of colouring on maps (especially since part of the Homann Cartographic Bureau had been transferred to Göttingen in 1755);[43] or even an attempt to reduce art to mathematical terms. Judging from Mayer's own remarks on the properties and advantages of making oil-paintings in accordance with his new art,[44] however, the benefits of its application to the reproduction of works of art in general and maps in particular seem to have been uppermost in his mind. Unfortunately, the treatise which he is known to have written on this subject is lost.[45] The motivation for his investigation of the theory of the magnet and the application of his results to the calculation of the variation and dip of the magnetic needle, was very likely due to his recognition—following his reading of Euler's *Dissertatio de magnete* (Paris, 1747), which impressed him greatly—that these phenomena might still prove to be a reliable means of determining geographical latitudes and longitudes.

Mayer's still-unpublished *Theoria Magnetis* (1760),[46] apart from containing a convincing verification of the inverse-square law of magnetism based more soundly on both theory and experiment than Augustin Coulomb's well-known researches with his torsion-balance a quarter of a century later,[47] is of interest in revealing (more clearly than any of his published works) the scientific attitude which Mayer was consciously adopting in the latter years of his life. In the introductory section, he examines the Cartesian view held by the majority of his contemporaries that magnetism is produced by a vortex of some rare form of matter in perpetual motion, and rejects the vortex hypothesis in general as being either false or at least useless and inept. Later on, he reveals his adherence to Newton's point of view when he remarks:

'The phenomena of universal gravitation, elasticity and hardness of bodies have been explained, although their causes remain mysteries; and they could not be better or more simply explained even if their ultimate causes were known. Indeed, Nature would appear to have hidden these causes for the very reason that knowledge of them serves no useful purpose—thus reminding us once again of the dictum that Nature does nothing in vain.'[48]

Taken in this context, Mayer's emphasis on useful knowledge really amounts to an affirmation of his belief that only those speculations which are capable of experimental verification are of any value to the progress of science. This is essentially the attitude which Lichtenberg recognizes as being typical of Mayer's work, when he remarks:

'Someone once said about Tobias Mayer: he himself had not known, that he knew so much—and therein lies something [which is] certainly very true. This is the real way to progress in the world. The ordinary scholars approach science teleologically, and even already in the title are anticipating something that they still do not know. That is depressing. Mayer himself was always seeking, and everything which he learned was to him a need: thus he could develop his science further. Nowadays one learns just the opposite, and offers syntheses which one will never need, along with a lot of useless things, although they are nevertheless very ingenious. Franklin seems to me to have been a similar [kind of] scholar. Meister had much of this attitude; also Cook. The last-mentioned said: 'To Hell with all scholarship', and [yet] he thought, learned, and studied thoroughly and was probably a greater scholar than many people whom he and the whole world called such. There is certainly also a great deal of truth in this distinction. The scholar could be someone who has stored a large amount of knowledge in his head, which may be of no further use to him except that he is able to communicate it again. If, however, someone becomes a specialist in a single subject, and all mankind agrees with him—and [if] he is a man only inasmuch as he is this, then he is no scholar.'[49]

This interesting passage could be interpreted as an oblique attack upon the scholastic mode of thought still current in Germany during the latter half of the eighteenth century, which had been sustained by the philosophical writings of Gottfried Wilhelm Leibniz and Christian von Wolff. The main point which Lichtenberg is wanting to make, however, is that: 'The major difference between English and German scholars is not so much the latter's preoccupation with the ancients, as the circumstance that the former are quickly made to understand thoroughly whatever they learn. They [the English] are not so easily satisfied and set more store on clear ideas. Our youth is spoiled by the awful mixture of reading, and nothing in the world is worse than our poets who are so much overflowing with sentiment.'[50] Mayer, Franklin, Meister and Cook were all men who were completely unbound to any philosophical tradition but shared the common ability to respond successfully to each intellectual challenge as it presented itself, without being side-tracked by irrelevant issues. Lichtenberg's recognition of Mayer's characteristically hypothetico-deductive experimental approach to science is of particular significance since it was derived from his close acquaintance with the latter's unpublished writings.

Before going on to enquire why Lichtenberg, rather than anyone else, came to be associated with editing some of these, it is necessary to examine certain events which occurred in the two decades preceding the original publication of this volume. After his financial circumstances had been ameliorated in 1754, Mayer had entered into a contract to buy a large house in Göttingen not far from his observatory, and received a generous loan for that purpose from the Hanoverian government.[51] However, owing to the very difficult conditions which were soon to persist after the French occupation of Göttingen during the Seven Years' War, he fell behind in his six-monthly repayments; and at the time of his death from gangrene on 20 February 1762, half the price (1400 Reichsthaler) had still to be repaid.[52] Now that her source of income had been cut off, his widow had no hope of ever maintaining the house—let alone settling this debt—unless the British Admiralty could be persuaded to advance some of the reward to which her late husband might be entitled, either in the form of a lump sum or as an annual pension. Michaelis, unaware of the seriousness of the situation, had written to Frau Mayer a few weeks beforehand promising that he would continue to protect her interests in this affair after her husband's death—a promise which happened to coincide with Mayer's own wish.[53] Michaelis had asked Kästner to question Mayer about what he would like to do regarding the longitude prize but, after trying unsuccessfully to obtain this information, Kästner wrote down six questions which he wanted answered and left them with Frau Mayer to put to her husband when he was in the right frame of mind to consider them. This was duly done, and apart from his wish that a Memorial should be sent after his death to the British Admiralty, Mayer expressed the desire that his solar and lunar tables should be published whether his heirs were awarded a prize or not, and that James Bradley should be informed of his death. All scientific enquiries concerning Mayer's discoveries were to be addressed to his brother-in-law Georg Moritz Lowitz, who had been his colleague in Nuremberg and had followed him as Ordinary Professor of 'Welt Weisheit und practischer mathematischer Wissenschaften' (Cosmography and practical mathematics) in Göttingen in February 1755.

A seventh item appended by Frau Mayer to Kästner's list at her husband's request was a statement to the effect that if the British Parliament were to award £10 000 to Mayer's heirs, £2000 should be gifted to the Göttingen Scientific Society, £1000 to Michaelis and £1000 to W. P. Best in London.[54] Michaelis, who was at that time the Director of the Göttingen Scientific Society, was informed of this intention over two weeks before Mayer's death;[55] and this put him under an even stronger moral obligation to extricate the latter's widow and her (at that time four) surviving male children from their financial predicament. He lost no time in writing to Best, informing him of Frau Mayer's difficult circumstances and suggesting that the Admiralty might be persuaded to grant her a sum to tide her over her immediate difficulties, or an annual pension of £50 or more.[56] Best, who now also stood to gain from a settlement of Mayer's claim, made a number of enquiries in London. Acting on the advice of the President of the Royal Society, he wrote a petition to the Board of Longitude requesting recognition for Mayer's heirs and urging that the required observations and experiments to test his discovery would soon be completed.[57] Unfortunately, however, the president of that Board, Lord Anson, had taken ill and had gone to Bath to recuperate, and there was now a new secretary of both the Admiralty and the Board, Sir Philip Stephens, who was not entirely familiar with the events which had already occurred. Moreover, the Admiralty was by this time fully committed to assessing the relative merits of Harrison's watch before a decision could be made regarding Mayer's lunar tables. To make matters worse, the Seven Years' War was still in progress and the Admiralty was occupied with affairs of national importance. Meanwhile, Frau Mayer's personal appeal to the Hanoverian government for an annual pension of 200 Reichsthaler, which she promised to refund if and when she should receive an award from England, also proved to be of little avail—not because of any lack of sympathy or goodwill but simply because money was already heavily overdrawn from the Hanoverian treasury as a result of the war.[58]

On account of the difficult conditions in England, and the fear of spoiling the chances of a favourable decision being reached on the question of the award, Best diplomatically did not press for it too enthusiastically at the wrong time but kept himself informed of what was going on.

Frau Mayer, however, blamed Michaelis for not doing enough to forward her claim and, acting upon advice which she received from De La Lande who was an official adjudicator in this matter, pressurized him into transmitting to Best a Memorial to the Admiralty which was duly referred to the Board of Longitude.[59] The result was that 'two gentlemen well-skilled in astronomy' (Nevil Maskelyne and John Robison) about to leave for the Barbados on H.M.S. Princess Louisa in order to test for a second time Harrison's timekeeper, were authorized to observe eclipses of as many of Jupiter's satellites as may be visible, and lunar distances, in order to compare the longitude determinations by these two methods and thereby test the accuracy of Mayer's lunar tables.[59] Robison, however, was obliged on account of family circumstances to withdraw his offer to undertake this mission and Maskelyne's assistant Charles Green agreed to go in his place.[60] The tables employed on this occasion were not the same as those used by Maskelyne during his voyage to St Helena in 1761, which constituted the basis of the lunar distances observations published in *The British Mariner's Guide* (London, 1763), but a new set which Mayer had gradually been improving during the seven years prior to his death, sent at Frau Mayer's request by Michaelis to Best in the summer of 1763, in accordance with Mayer's dying wish.[61] Frau Mayer also persuaded Michaelis to prepare a Memorial for her to copy in her own hand, which Best received from Michaelis less than two months later and duly transmitted to Sir Philip Stephens.[62]

This was the state of affairs when she received from Nuremberg copper-engravings of six segments of a lunar map.[63] The entire map, the drawings of which her late husband had managed to complete during his years in Göttingen, consisted of 14 segments—two being of the portion revealed by the Moon's libratory motion—suitable for pasting on to a spherical surface, since Mayer's intention had been to prepare a lunar globe.[64] In accordance with Kästner's advice, Frau Mayer wrote to von Munchhausen on 21 September 1763 offering the Hanoverian government the first opportunity to purchase these copper-plates and drawings for the Göttingen Observatory which was now being directed by Kästner himself. Unknown to her, the latter also wrote to Hanover recommending that a competent person should be found who could complete this fine piece of work and submit it to a printer.[65] The matter was then brought to the attention of King George III who asked Michaelis (through the usual diplomatic channels) if any of the Göttingen professors could undertake this. Before the end of the year, the copper-plates, drawings, *and the numerous other unpublished manuscripts* belonging to Mayer were purchased from his widow by the Hanoverian government for the Göttingen Observatory for the sum of 200 Reichsthaler, somewhat less than that which she was hoping to receive but all that the treasury was then able to afford.[66] There the matter rested for the next ten years. By then certain circumstances which will be related presently caused a revival of interest in the publication of Mayer's unpublished manuscripts in general, and the segments of the planispheres for the lunar globe in particular, which led to the original appearance of the *Opera inedita* . . .

Frau Mayer's receipt of this small sum of money was a godsend to her as a temporary, or interim, solution of her financial embarrassment, since she could not expect the commissioners of longitude to come to any decision before the results of the observations made by Maskelyne and Green during the sea voyage to Barbados had been properly attested and the respective merits of the lunar distances and timekeeper methods of longitude determination duly assessed. Towards the end of 1764 von Munchhausen was using his informal but influential contacts in London to make private representations on behalf of the widow and children of 'such an eminent astronomer as the late Professor Mayer was',[67] while Best also ensured that the matter would be kept alive in the minds of the commissioners of longitude by writing to remind them of the background to Mayer's claim and the urgency of a positive settlement.[68] The long-awaited decision was finally made at a memorable meeting of the Board of Longitude on 9 February 1765, at which both the recommendations affecting the Parliamentary rewards to Harrison and to Mayer were announced.[69] In a nutshell, the Board's attitude seems to have been that Harrison's watch was a practicable but not generally useful solution of the longitude problem, whereas the method based on Mayer's lunar tables was generally useful but not in itself practicable; and the terms of the 1714 Act required the solution to be both generally useful *and* practicable. At any rate, this interpretation would explain why the Parliament then awarded to Harrison £10 000 although his famous watch had clearly demonstrated its ability to keep time to well within the limit of accuracy required to find longitude at sea to within ½°, for which the appropriate reward should have been £20 000; and why Mayer's heirs were to receive a sum 'not exceeding £5000' for tables which, as Nevil Maskelyne claimed in a memorial presented at that very same meeting, were generally reliable as a basis for the determination of longitude at sea to within 1°. The appropriate reward in this case should have been £10 000—the sum which Mayer himself had anticipated receiving. Three months later Parliament's decision to accept the Board's recommendation regarding Harrison and to award Frau Mayer only £3000 for her late husband's tables, was a great disappointment to both parties concerned. The reduction of the latter reward and the decision to give Leonhard Euler an unsolicited prize of £300 in view of Mayer's analytical method, were undoubtedly consequences of a public protestation by Alexis Clairaut in *The Gentleman's Magazine* for May 1765. This supported his own and Euler's right for recognition in laying the foundations of the theory of the Moon's motion. Thanks to the intervention of no less a personage than King George III himself, Harrison was ultimately successful (in 1773) in receiving the second half of the reward to which he considered himself entitled; by this time, however, copies of his masterpiece had been made by independent craftsmen and found to keep time with a comparable degree of precision, thereby proving the general utility of the *chronometer* method (as it was now being called).[70]

All that was lacking, in 1765, respecting the method which Mayer had been championing, was a means of reducing the complexity and length of the computations involved, which were taking 'a few hours not exceeding four hours' to complete.[71] The Hadley octant, or the nautical sextant (which owes its origin partly to Captain John Campbell's unsuccessful experiments with Mayer's repeating circle) were already capable of yielding an angular measurement to an accuracy of well within 10′ in the hands of an experienced navigator, and this was adequate for observing lunar distances.[72] The Astronomer Royal, in his memorial, proposed that the practical application of this method could be facilitated by the preparation of a nautical ephemeris containing tables of lunar distances calculated from Mayer's 'last manuscript Lunar Tables', and additional astronomical tables to correct the observations for the effects of refraction, parallax, etc.—a suggestion which the Board immediately approved. Maskelyne's plan for the ephemeris was accepted by the Board at its meeting on 30 May 1765, and crystallized less than two years later into the *Nautical Almanac for 1767*. This made the mariner's calculations simpler and shorter, and thereby removed the only obstacle that had hitherto prevented the lunar distances method of finding longitude at sea from being considered as practicable. In his preface, written on 12 December 1766, Maskelyne states that:

'All The Calculations of the Ephemeris relating to The Sun and Moon were made from Mr. Mayer's last manuscript Tables, received by the Board of Longitude after his Decease, which have been printed under my Inspection and will be published shortly'.

Indeed, the lunar theory appeared during the following year,[73] while the edited version of Mayer's solar and lunar tables was published in Latin and in English in 1770.[74] Thus, by providing the first reliable basis for finding longitude at sea, Mayer's tables played a vital and fundamental role in oceanic navigation for several decades before the rival chronometer method (which ultimately superseded it) came into widespread use. It was to this that Robert Grant was referring in his *History of Physical Astronomy* (London, 1852), p. 488, when he remarked:

'Mayer has earned for himself an imperishable reputation by his lunar tables. Some astronomers may have achieved results of greater brilliancy, but few have been privileged to contribute so directly by their labours to the well-being of their fellow-men.'

Plate 4. Georg Christoph Lichtenberg

Georg Christoph Lichtenberg (1742–1799) and the editing of Tobias Mayer's unpublished works

Georg Christoph Lichtenberg (Fig. 4, Plate 4)[75] was born on 1 July 1742 in the village of Oberramstadt near Darmstadt, the youngest of a family of 18 children, of whom only one sister and four brothers survived childhood. His father, Joseph Konrad Lichtenberg, was the local pastor who used to take pleasure in introducing popular expositions of topics in astronomy and natural philosophy into his sermons. During the week, he gave lessons in mathematics and natural science to his children. Georg, who inherited many of his father's secular interests, benefited from this instruction until the time of his father's death in 1751. He was sent that same year to the high school in Darmstadt, where he revealed his aptitude for mathematics, physics, astronomy and astrology; and also helped to coach some of his fellow pupils in mathematics. In 1763, he obtained a grant from the Duke of Hessen-Darmstadt to enable him to study at the Georg-August Academy in Göttingen, which had just entered a phase of expansion following the years of great hardship during the Seven Years' War.[76] During the next four years, he studied philosophy and mathematics, physics, history and philology, and read a number of English humorous novels for his amusement and pleasure. He suffered a period of depression following the death of his mother during his second year as a student, which was aggravated by his sensibility to his physical deformity—he was a hunchback—of which he had now become acutely aware; nevertheless, he completed his studies successfully in the spring of 1767.

By now, he was finding himself increasingly attracted towards astronomy, and fortunate enough to be able to prolong his higher education in Göttingen with financial support from the Minister of State for his native Hesse (von Riedesel) on the understanding that he would be called to a professorship at Giessen as soon as circumstances there should permit. In order to augment this income, Lichtenberg acted as a tutor to several English boys who had come to the Georg-August Academy, which was then regarded as being one of the best universities in Europe, to complete their education. The father of one of these boys, Lord Boston, invited Lichtenberg in 1770 to accompany his son (William Irby) to England at the end of the spring semester and to spend a few weeks at his country home, a lovely mansion on the outskirts of London. During this visit, the first of two which Lichtenberg made to the country whose literature he so much admired,[77] he was introduced by his host to a number of distinguished personages, including King George III himself whom he met at the Court Observatory in Richmond. The King already knew of Lichtenberg through two reports which had appeared in the *Göttingische*

Anzeigen . . . concerning the 1767 earthquake and the 1769 transit of Venus;[78] but it was more likely to have been this personal introduction and George's policy of attracting Englishmen and foreigners to study at Göttingen, and not Lichtenberg's reputation as an astronomer, which persuaded him to instruct his government in Hanover soon afterwards to offer Lichtenberg an Extra-Ordinary professorship in philosophy at the Georg-August Academy.[79]

Towards the end of 1771 Lichtenberg, on the recommendation of his tutor Kästner (with whom he was at that time still on friendly terms), was commissioned to begin an accurate determination of the geographical co-ordinates of a number of towns in North-West Germany which George III required as the basis of a reliable atlas of that region for his military engineers. Since the geographical co-ordinates of Göttingen had already been found with sufficient accuracy by Mayer himself, Hanover was the first town in which he was ordered to begin his measurements. There was, however, no instrument in either Göttingen or Hanover which would have been suitable for making the necessary astronomical observations, so the king ordered an accurate quadrant to be made by Jeremiah Sisson in London specifically for this purpose. This instrument was duly dispatched to Hanover, where Lichtenberg ruled and tested it before commencing his programme of meridian altitude observations on 26 March 1772. He completed this work on 13 July and returned to Göttingen where, less than a month later, he learnt that the King wanted him to undertake similar measurements at Osnabrück and Stade. He carried these out and completed them by December of the following year.[80] The results, published in 1778,[81] were generally considered to be among the best of their kind then in existence—a fact which can be confirmed by comparing them with those obtained from similar measurements made half a century later by Karl Friedrich Gauss and Heinrich Christian Schumacher.[82] On the other hand, they suffer from the fact that Lichtenberg failed to specify the precise location of his temporary observatories, the uncertainty due to this cause being greatest in the case of Osnabrück. Besides his quadrant, which sufficed for the latitude determinations, Lichtenberg employed a precision-clock and an achromatic telescope of six-times magnification to time the eclipses of the Moon and Jupiter's satellites from which he deduced the longitudes. He also measured the variation of a magnetic needle at each of his three observing stations.[13]

This work, though not remarkable for its novelty, is important in the present context since it made Lichtenberg appreciate the full significance of Mayer's pioneering attempts to improve the accuracy of geodetic and astronomical instruments. In the introduction to his report on the results of his investigations, published in the seventh volume of the new transactions of the Göttingen Scientific Society (1776), he discusses in detail the instrumental errors in his quadrant, and remarks:

'The entire investigation [of these errors] follows the tradition of Tobias Mayer, who was the first in Germany—about the same time as Bradley in England—to have made thorough determinations of the instrumental errors and the errors of eccentricity in ruling, and also to have developed methods of reducing the error to a minimum.'[84]

The last phrase is an allusion to the repeating principle (*artificium multiplicationis*) applied by Mayer to his new instrument for angle-measurement in the Göttingen Transactions for 1753,[85] which is also the basic feature of the Repeating Circle described and illustrated for the first time in Maskelyne's edition of Mayer's solar and lunar tables (1770).[86] It is highly probable, therefore, that Lichtenberg's close contact with this aspect of Mayer's work caused him to decide, ultimately, to incorporate the latter's unpublished description of his experiments on John Bird's mural quadrant into the *Opera inedita* (cf. II).

It was in the course of his short visit to Hanover on his return journey to Göttingen from Osnabrück after completing his observations there in January 1773, that Lichtenberg first broached the subject of the lunar drawings and manuscripts of Mayer which were still being allowed to lie neglected in the Observatory and at the Royal Society of Sciences in Göttingen—despite Kästner's official recommendation almost ten years previously that someone competent should be found to prepare them for publication.[87] As a result of informal discussions with several important people in Hanover (one of whom was very likely to have been Christian Schernhagen, a Councillor at the Royal Chancery), Lichtenberg was asked to submit a written report on the matter to the authorities there on the understanding that his name would not feature in any acrimonious discussions which might ensue. Thus, shortly after his return to Göttingen, he wrote a memorial to the Hanoverian government stating that he himself would be willing to edit the papers in question.[88] In it, he refers to Johann Heinrich Lambert's regret that Mayer's works were not known, and claims that Maskelyne (whom he probably met during his visit to England) had assured him that Mayer's lunar map and indeed all his manuscript writings would willingly be published in London by the Commissioners of Longitude.[89] Unpublished items to which he explicitly refers are:

(1) A large number of observations, and a well-drawn lunar map.
(2) A catalogue of zodiacal stars, which seemed to be as complete as all previous lists published in England and France.
(3) Tables for the Moon's motion.
(4) A new method of computing solar eclipses.
(5) A new theory of the magnet.

(6) A new explanation of an earthquake.

(7) A treatise on the measurement of colour.

Of these, (2) and (3) were treatises which he particularly wanted to see edited and published, if a publisher could be found in Göttingen; (5), (6) and (7) were still unknown in foreign countries, although short abstracts concerning them had appeared—while Mayer was still alive—in the *Göttinger gelehrte Anzeigen.*[90]

The contents of this memorial were duly communicated to George III, who authorized von Munchhausen to instruct Kästner and the Royal Society of Sciences in Göttingen to allow Lichtenberg to have access to the papers in their custody. This was done on 1 April 1773[91] and the documents were all in Lichtenberg's possession only four days later. This gave him several days to examine them carefully before setting out on his journey to Stade; also, to engage the services of the local artist Joel Paul Kaltenhofer, and the printer Johann Christian Dieterich with whom he was already intimately acquainted. He commissioned Kaltenhofer to make an accurate copper-engraving of the smaller copy of Mayer's two well-preserved lunar maps, since this was apparently in accordance with yet another of Mayer's dying wishes. Besides the obvious advantages of his living in Göttingen, Kaltenhofer was generally regarded as one of the best engravers in Germany at that time. Moreover, he had been a personal friend of Mayer's and had himself made telescopic observations and drawings of the Moon.[92] During the summer months, while Lichtenberg was in Stade, Kaltenhofer finished two trial engravings of Mayer's lunar map. The first, done without any specific instructions, did not please Lichtenberg at all[93]—among other faults the contrast in shading was too sharp. The second seemed more like the original drawing and met with his approval,[94] although he was later to state a clear preference for the excellent engraving made, from the same drawing, by Georg Heinrich Tischbein for Johann Hieoronymus Schröter's *Selenotopographische Fragmente* (Lilienthal, 1797).[95] On Lichtenberg's instructions, Kaltenhofer had superimposed a grid from which one can read off the values of the selenographic co-ordinates of the lunar markings.[96]

As a result of his own perusal of Mayer's writings, and his discussions and correspondence with Dieterich, Lichtenberg had made up his mind by September 1773 to omit those numbered (3), (5), (6) in his memorial, and to include instead those on the variations of a thermometer (cf. I) and on the proper motion of stars (cf. VI).[97] The lunar tables turned out to be just the early drafts of those published in the Göttingen *Commentarii* . . . for 1753, which in turn had been superseded by the 'new improved' tables published by Maskelyne in 1770. The theory of the magnet was much longer than any of the other treatises, and bore little resemblance to them either in content or in style. Dieterich may well have been reluctant to publish it in any case on account of the numerous mathematical symbols and formulae which it contains. Mayer's attempt to explain the origin of an earthquake had been published previously, in 1756,[98] and it is apparent from a document preserved in the archives of the Göttingen Academy of Sciences, dated 24 October 1786, that Lichtenberg had a very poor opinion of its intrinsic merit.[99].

The primary consideration affecting Lichtenberg's decision to include Mayer's treatise on the mural quadrant and the other two already mentioned (viz. I and VI), was the ease with which these could be edited. As he admits:

'The complete lectures read before the Society fill a magnificent volume, the editing of which was comparatively simple and quickly done. A second volume should contain the fragments with explanatory notes taken from the abstracts [in the *Gött. gel. Anzeigen*] and both, in accordance with the plans of the publisher Dieterich, should constitute the conclusion of a complete edition of Mayer's works. The printing of the second volume, like the materialisation of this complete project, are still pending today.'[100]

Nor was it ever begun, for Lichtenberg had lost his initial enthusiasm for this work.[101] Only his strong moral obligation to Mayer's heirs, to the Hanoverian government, and to men like Johann Heinrich Lambert, Joseph De Lalande, and the ex-Jesuit Maximilian Hell, to do justice to the memory of the man whom they so greatly admired, prevented him from yielding to the temptation of making the short journey to Hamburg and embarking upon the next ship bound for England—the 'happy island' to which he longed so much to return.[102]

Two letters to Lambert on 1 March and 3 April 1774 reveal the progress which he had been making during the three months following his return to Göttingen at the end of the previous year;[103] and in particular, his problems in editing Mayer's treatise on the affinity of colours (cf. IV), to which he was to devote the longest commentary. Paul Hahn implies that Lichtenberg's concern with the two physical—as opposed to astronomical—treatises in the *Opera inedita* . . . (viz. I and IV) was due to the fact that the editor happened to be a physicist himself,[104] but this seems to have no significance except inasmuch as Lichtenberg considered the observation of nature to provide the objective truth which, when combined with self-observation or subjective truth, would reveal the true nature of reality.[105] At the end of April 1774, he requested the Hanoverian government's permission to have a paid holiday in England for the months of June and July;[106] this was granted, but his duties in connection with his editing of the *Opera inedita* caused him to postpone this second journey to London until 24 August. Lichtenberg's mood is reflected in a letter to his friend Dieterich written not long after he arrived there, where he says:

'I hope that all has gone well with the printing, though I am now far from content with the whole thing, and would gladly be quit of it all. The preface and dedication were begun in Hanover, continued on the Schützenkrug behind Rehburg, and concluded in Osnabrück. Heaven knows how it has turned out.'[107]

Since Lichtenberg had by then composed his Latin commentaries, and written the remarks on Mayer's selenographic table which are appended to this work, it is likely that his concern must have been for the quality of his Latin rather than for the clarity of his two-page summary of the contents.[108] If the 'it' in the last sentence of the above quotation refers to his editorial work as a whole, one can but reply that it was not very thorough. Lichtenberg seems not to have understood the method which Mayer followed for computing solar eclipses (cf. III), otherwise he would surely have attempted to correct several errors which it contains; nor did he notice a fairly obvious algebraic error incurred by Mayer in his analysis of colour-mixing (cf. IV) which invalidated one of his conclusions. It is hoped that the following additional commentaries will compensate adequately for these particular deficiencies, and help to remove a number of obscurities in Mayer's writings, which Lichtenberg himself failed to perceive or, at least, to clarify.

Translator's notes on Mayer's treatises

I. *A more accurate definition of the variations of a thermometer* (1755)

The most interesting feature of this treatise lies in the method rather than in the content; or, as Delambre puts it, in the ingenuity rather than in the validity of its concepts.[109] Mayer advocates the application of the empirical approach, which he and other astronomers had adopted so successfully in the description of the lunar and planetary motions, to what he sees as an analogous problem of describing annual and daily temperature inhomogeneities which depend on the latitude, altitude and climatic conditions of any given place. Lichtenberg's commentary, on the other hand, is less concerned with the method than with the errors which arise in practice in the graduation of thermometers. He describes the thermometer made by Mayer in 1755, which was used in connection with the latter's subsequent astronomical observations; and draws attention to a small systematic discrepancy of 1/62 000 on the Réaumur scale[264] (one of six scales in which this thermometer was graduated). The error of +0·2% which was thereby incurred by Mayer in his determinations of astronomical refraction proved, however, negligible in practice.

II. *Astronomical observations taken on the mural quadrant in the observatory at Göttingen* (1756)

This treatise contains Mayer's formula for the meridian transits of stars, but not his proof of that formula. The errors in levelling the quadrant, setting it in the correct azimuthal direction (i.e. in the plane of the meridian), and in assuring that the viewing telescope is properly collimated (i.e. with its axis parallel to the plane of the quadrant), are each considered by Mayer in turn. He states that the

$$\text{levelling error} = \frac{A \cos \alpha}{\cos \delta}$$

$$\text{azimuthal error} = \frac{B \sin \alpha}{\cos \delta}$$

$$\text{collimation error} = \frac{C}{\cos \delta}$$

A, B, C being constants whose values can in each case be determined by a single observation, α the zenith distance of a star or other heavenly body, and δ its declination north or south of the equator. The total error is then taken to be the sum of the individual terms; namely

$$\frac{A \cos \alpha + B \sin \alpha + C}{\cos \delta}.$$

This manner of presentation suggests that Mayer's skill in geometrical projection enabled him to arrive at the expression for each of the three independent errors directly, without having recourse to calculation. For the benefit of the modern reader, however, a rigorous analytic proof based upon the conventional method of spherical trigonometry, is given below.

In Fig. 5, which applies to observations made when the star X is at its upper transit, let us suppose that small errors, A and B radians, exist in the altitude and azimuth (measured with the mural quadrant), respectively, due to the inaccuracy of aligning that instrument, and that,

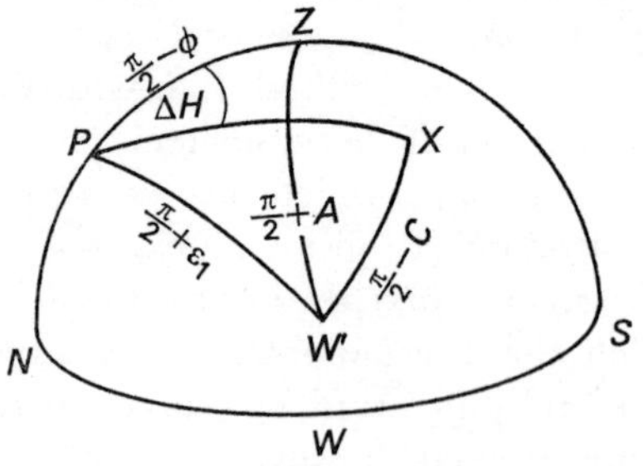

Fig. 5

owing to a deficiency in the latter's construction, the axis of rotation of the viewing telescope is inclined at an angle of slightly less than $\pi/2$ radians ($=\pi/2-C$, say) to the optical axis itself. The combined effect of these three errors will be to produce an apparent change in the local hour angle (from zero to ΔH, say), and to displace the point of intersection of the axis of the instrument on the celestial sphere from W to some neighbouring point W'. Thus we get (Fig. 5):

$$ZW' = \frac{\pi}{2} + A$$

$$S\widehat{Z}W' = \frac{\pi}{2} + B$$

$$XW' = \frac{\pi}{2} - C$$

Let

$$PW' = \frac{\pi}{2} + \epsilon_1$$

$$ZPW' = \frac{\pi}{2} + \epsilon_2$$

where ϵ_1, ϵ_2 are small quantities which will be eliminated in the following analysis.

Applying the cosine formula[110] to the spherical triangle PXW':

$$\cos XW' = \cos PW' \cos PX + \sin PW' \sin PX \cos \widehat{XPW'}$$

$$\sin C = -\sin \epsilon_1 \sin \delta + \cos \epsilon_1 \cos \delta \cos \left(\frac{\pi}{2} + \epsilon_2 - \Delta H\right)$$

$$\sin (\Delta H - \epsilon_2) = \frac{\sin C + \sin \epsilon_1 \sin \delta}{\cos \epsilon_1 \cos \delta}$$

To a sufficient approximation, since C, ϵ_1, ϵ_2 are regarded as being very small quantities, one may therefore write:

$$\Delta H = \epsilon_2 + \frac{C + \epsilon_1 \sin \delta}{\cos \delta} \tag{3}$$

Applying (1) to the spherical triangle PZW':

$$\cos PW' = \cos PZ \cos ZW' + \sin ZW' \cos PZW'$$

$$-\sin \epsilon_1 = -\sin \phi \sin A + \cos \phi \cos A \cos \left(\pi - \frac{\pi}{2} + B\right)$$

$$\sin \epsilon_1 = \sin \phi \sin A - \cos \phi \cos A \sin B$$

Approximating,

$$\epsilon_1 = A \sin \phi - B \cos \phi \tag{4}$$

Similarly,

$$\cos ZW' = \cos PZ \cos PW' + \sin PZ \sin PW' \cos ZPW'$$

$$-\sin A = -\sin \phi \sin \epsilon_1 - \cos \phi \cos \epsilon_1 \sin \epsilon_2$$

Approximating,

$$A = \epsilon_1 \sin \phi + \epsilon_2 \cos \phi \tag{5}$$

Eliminating ϵ_1 from (4) and (5):

$$A = (A \sin \phi - B \cos \phi) \sin \phi + \epsilon_2 \cos \phi$$

$$\therefore \epsilon_2 = \frac{A - A \sin^2 \phi + B \cos \phi \sin \phi}{\cos \phi} = A \cos \phi + B \sin \phi \tag{6}$$

Eliminating ϵ_1 and ϵ_2 by substituting (4) and (6) into (3), one obtains:

$$\Delta H = (A \cos \phi + B \sin \phi) + \frac{C + (A \sin \phi - B \cos \phi) \sin \delta}{\cos \delta}$$

$$= \frac{A(\cos \phi \cos \delta + \sin \phi \sin \delta) + B(\sin \phi \cos \delta - \cos \phi \sin \delta) + C}{\cos \delta}$$

$$= \frac{A \cos (\phi - \delta) + B \sin (\phi - \delta) + C}{\cos \delta}$$

At upper transit, $\phi-\delta=\alpha$, the zenith distance of the star. Thus we finally obtain

$$\Delta H=\frac{A\cos\alpha+B\sin\alpha+C}{\cos\delta}$$

which is Mayer's formula for the total error in the time of upper transit, due to the levelling, azimuthal and collimation errors involved in the adjustment and construction of the mural quadrant. A similar formula can easily be obtained for lower transit.

An article written earlier this century by G. van Biesbroeck[111] draws attention to an interesting similarity between this formula and a rule which the Danish astronomer Olaus Roemer (1644–1710) effectively applied to obtain precisely the same corrections for meridian transit observations of stars made with his *Rota meridiana*[112] several decades beforehand. This being so, it would appear that Roemer could be remembered as having prior claim to the discovery of this formula; consequently *his* name, and not Mayer's, should have been associated with it. This has been further corroborated by Elis Strömgren,[113] who also drew attention to the fact that Mayer actually computed the errors in Roemer's instrument by means of the formula in question. Mayer's reason for doing so was that he wanted to check Roemer's reductions of meridian transit observations made between 23 and 26 October 1706 before employing them in his study of the proper motions of stars (cf. VI, p. 109, also note 252). The only known source of information regarding Roemer's own treatment of the instrumental errors is contained among a collection of notes in a portfolio bearing the title *Adversaria* which escaped destruction and was transmitted by Roemer's widow in 1739 to the University of Copenhagen. Since the *Adversaria* were not published until 1910,[114] van Biesbroeck and Strömgren are merely questioning Mayer's priority, and not accusing him of plagiarizing Roemer's results. Indeed, there is no indication in any of Mayer's published or unpublished works, or in his preserved correspondence, that he could possibly have had access to such knowledge before composing his lecture on this subject in 1756. Thus, he would still appear to deserve credit for originality.

III. *A simple and accurate method of computing solar eclipses visible in a given place* (1757)

Lichtenberg's failure to supply a commentary to this treatise is inexcusable; either he did not fully understand the method which Mayer outlines in it or, by the time he came to examine it, he was already in a hurry to complete his editorial task, in order to be able to set off on his second journey to England. The original text has been amended in the English translation to remove the obscurities which are incurred by Mayer on account of:

(i) The ambiguous symbol μ (cf. notes 119 and 209) and an error regarding its sign which may have been made by Mayer himself, by Lichtenberg in his transcription, or by Dieterich in printing.

(ii) The omission of brackets in the formula for the increase of the Moon's semi-diameter (cf. p. 74).

(iii) The lack of any explanation of the manner by which the tabulated corrections for latitude and equatorial parallax were obtained.

(iv) A typographical error in which π is introduced instead of ϕ in two of the three equations cited in the fourth precept. This makes nonsense of the definition of the angle ω, which should really be given by $\tan\omega = \tan\phi/\sin\mu$ (cf. p. 76).

(v) The occurrence of the second term in the equation for the parallax in latitude (cf. p. 76), whose value ought to be zero under the assumption by which the corresponding parallax in longitude (on which it is dependent) was obtained. This will be justified presently.

The following mathematical demonstration of the formulae which occur in the text should help to make Mayer's method more comprehensible and illuminate the precepts which he employs.

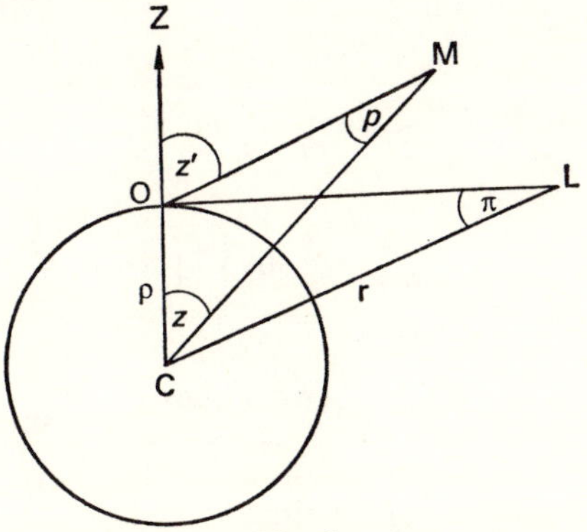

Fig. 6

Consider first the augmentation in the semi-diameter of the Moon which occurs as a result of that body being seen by an observer situated at some point O on the Earth's surface instead of at its geometrical centre C (cf. Fig. 6). Let M denote the position of the Moon;

$Z\widehat{O}M = z'$, the Moon's zenith distance, corrected for astronomical refraction;

$Z\widehat{C}M = z$, the Moon's geocentric zenith distance;

$O\widehat{M}C = p$, the Moon's parallax (at latitude ϕ).

Then, by definition, the

$$\text{augmentation} = \frac{\text{Increase in Moon's semi-diameter}}{\text{Moon's semi-diameter}} = \frac{\text{MC}-\text{MO}}{\text{MC}} = 1 - \frac{\text{MO}}{\text{MC}}$$

In $\triangle$MOC,

$$\frac{\text{MC}}{\sin \widehat{\text{COM}}} = \frac{\text{MO}}{\sin \widehat{\text{OCM}}} \quad \text{or} \quad \frac{\text{MO}}{\text{MC}} = \frac{\sin z}{\sin z'};$$

also,

$$p = z' - z \quad \text{or} \quad z' = z + p; \quad \text{and so} \quad \frac{z'+z}{2} = z + \frac{p}{2}.$$

Hence,

$$\text{augmentation} = 1 - \frac{\sin z}{\sin z'} = \frac{\sin z' - \sin z}{\sin z'}$$

$$= \frac{2 \sin \dfrac{z'-z}{2} \cos \dfrac{z'+z}{2}}{\sin z'}$$

$$= \frac{2 \sin \dfrac{p}{2} \cos \left(z + \dfrac{p}{2}\right)}{\sin z'} = \frac{2 \sin \dfrac{p}{2} \left(\cos z \cos \dfrac{p}{2} - \sin z \sin \dfrac{p}{2}\right)}{\sin z'}$$

Since

$$p \cong 1° = 3600'', \quad \sin \frac{p}{2} = \frac{p}{2} \sin 1'', \quad \cos \frac{p}{2} \cong 1;$$

thus

$$1 - \frac{\text{MO}}{\text{MC}} = \frac{2\dfrac{p}{2} \sin 1'' \left(\cos z - \dfrac{p}{2} \sin 1'' \sin z\right)}{\sin z'}$$

$$= \frac{p \sin 1'' \cos z}{\sin z'} - \frac{p^2}{2} \sin^2 1'' . \frac{\text{MO}}{\text{MC}}$$

$$1 - \frac{\text{MO}}{\text{MC}}\left(1 - \frac{p^2}{2} \sin^2 1''\right) = \frac{p \sin 1''}{\sin z'} \cos z$$

But

$$\frac{p^2}{2} \sin^2 1'' \cong \frac{(3600)^2}{2} \times \frac{1}{(206265)^2} \cong 1{\cdot}5 \times 10^{-4},$$

which is negligible compared with unity; so we may write

$$1 - \frac{\text{MO}}{\text{MC}} = \frac{p \sin 1''}{\sin z'} . \cos z$$

In Fig. 7, ϕ denotes the latitude of the observer; MR $= \delta$, the Moon's declination, or height above the equator; ZM $= z$, the Moon's geocentric

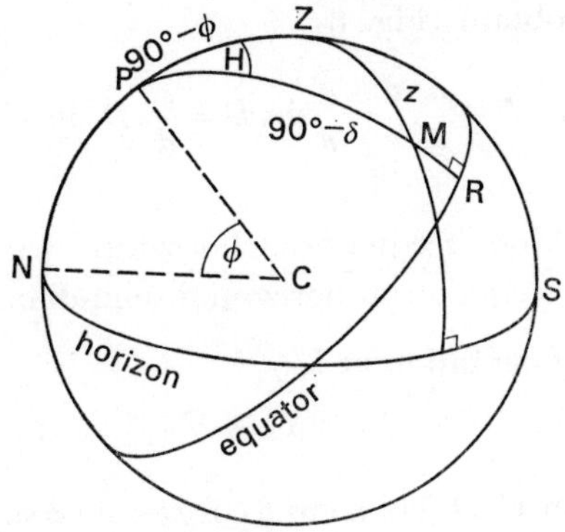

Fig. 7

zenith distance; and $\widehat{ZPM}=H$, the Moon's hour angle, which, by convention is measured westwards from the observer's meridian PZS.

Thus, applying the cosine formula to the spherical triangle ZPM, one immediately obtains:

$$\cos z = \cos (90° - \phi) \cos (90° - \delta) + \sin (90° - \phi) \sin (90° - \delta) \cos H$$

or

$$\cos z = \sin \phi \sin \delta + \cos \phi \cos \delta \cos H$$

Consequently, to a sufficient approximation:

$$\text{Augmentation} = 1 - \frac{\text{MO}}{\text{MC}} = \frac{p \sin 1''}{\sin z'} (\cos \phi \cos \delta \cos H + \sin \phi \sin \delta)$$

When the centre of the Moon's disc is on the horizon, at L (cf. Fig. 6), $z' = 90°$ and $p = \pi$, the Moon's horizontal parallax. Hence

$$\sin z' = 1, \quad \text{and} \quad \pi \sin 1'' = \sin \pi = \frac{\text{CO}}{\text{CL}} = \frac{\rho}{r}$$

say, where ρ is the radius at latitude ϕ of the non-spherical Earth, and r the Moon's geocentric distance.

Let a be the equatorial radius of the Earth; d the (linear) radius of the (spherical) Moon, and D the Moon's *equatorial* horizontal semi-

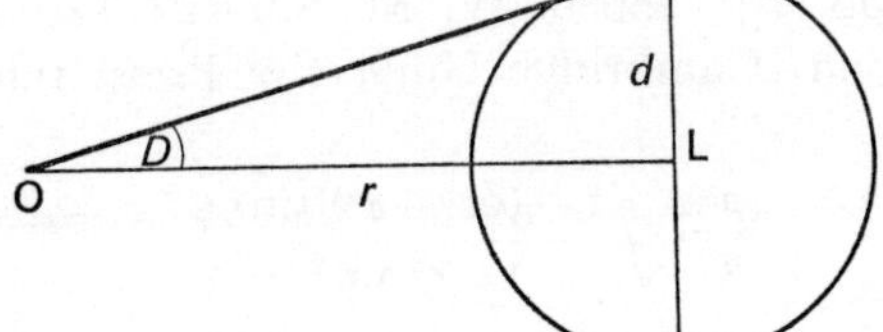

Fig. 8

diameter. Then we obtain (Fig. 8):

$$\pi \sin 1'' = \frac{a}{d} \cdot \frac{d}{r} = \frac{a}{d} \sin D = \frac{a}{d} . D \sin 1''$$

Then:

$$\frac{\text{Increase in Moon's equatorial horizontal semi-diameter}}{\text{Moon's equatorial horizontal semi-diameter}}$$

$$= \text{augmentation at L}$$

$$= \pi \sin 1'' (\cos \phi \cos \delta \cos H + \sin \phi \sin \delta)$$

$$= \frac{a}{d} \sin 1'' \, D \, (\cos \phi \cos \delta \cos H + \sin \phi \sin \delta)$$

Substituting the modern values of $a = 6378$ km and $d = 1738$ km into this expression, we obtain the result that the increase in the Moon's equatorial horizontal semi-diameter measured in seconds of arc is:

$$\frac{D^2(\cos \phi \cos \delta \cos H + \sin \phi \sin \delta)}{56208}$$

This is essentially the same formula quoted by Mayer (cf. p. 74), although D is now more precisely defined, and the brackets are included in the numerator as they should be. In Fig. 7, $H = \mu$, the distance measured along the equator from the observer's meridian; if, however, the Moon were situated to the east of that meridian, $H = 360° - \mu$; but the formula would remain unchanged. A $\pm$ in front of the second term would take account of the case where the Moon is situated to the south of the equator; that is, where it has a negative declination. The slight discrepancy of 0·08% between the numerical value of the denominators of the above equation and that given in the original text can be ascribed entirely to the fact that Mayer took $a/d = 11/3 = 3 \cdot 6667$ as compared with the modern value of $3 \cdot 6697$ for this ratio.

Mayer now considers the corrections which still require to be made to the quantities D and ϕ due to the fact that the Earth is not a perfect sphere. These are given for a spheroidal Earth by the following formulae, which do not appear in the text; but the accompanying figure (Fig. 9) should suffice to indicate the nature of the corrections to be applied. The proofs can be found on pp. 195–200 of W. M. Smart's *Text Book on Spherical Astronomy*, 5th edition (Cambridge University Press, 1962) or in similar works.

$$\frac{\rho}{a} = \sqrt{\frac{1 - (2e^2 - e^4) \sin^2 \phi}{1 - e^2 \sin^2 \phi}}$$

where $\rho = CO$ is the Earth's semi-diameter at latitude ϕ; and $e^2 = 1 - b^2/a^2$.

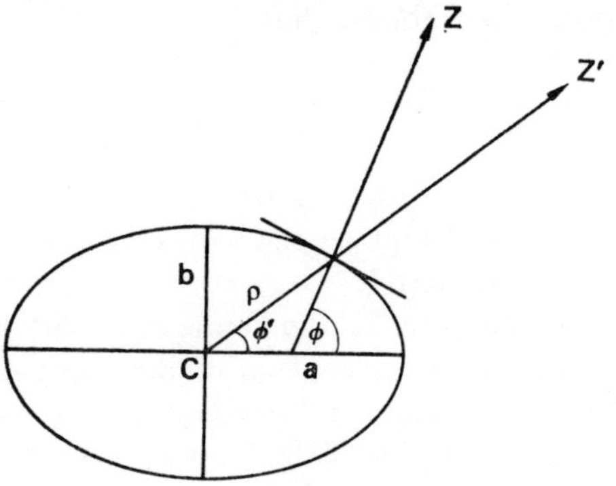

Fig. 9

The variation in latitude

$$\phi - \phi' = m\frac{\sin 2\phi}{\sin 1''} - \tfrac{1}{2}m^2\frac{\sin 4\phi}{\sin 1''} + \tfrac{1}{3}m^3\frac{\sin 6\phi}{\sin 1''} - \ldots$$

where

$$m = \frac{e^2}{2-e^2} = \frac{a^2-b^2}{a^2+b^2}$$

It is easy to verify with the aid of the latter equation that the values of the latitude corrections tabulated by Mayer (Table 8) correspond to a value of $1/200$ for the ellipticity of the Earth $(a-b)/a$; this is somewhat greater than the respective values of $1/220$ and $1/230$ which he adopted in his unpublished lectures on theoretical and spherical astronomy (1751)[115] and on mathematical geography (1760),[116] so it would appear that he was deliberately using a simpler fraction in order to facilitate his calculations. The maximum error incurred (for the latitude $\phi=45°$) is $5'\,24''$, and Mayer's tabulated value is greater than the actual variation of $11'\,47''$ by that amount. The relation between the horizontal parallax of the Moon observed at the terrestrial latitude ϕ (viz. π) and its equatorial horizontal parallax D is given by:

$$\sin \pi = \frac{\rho}{r} = \frac{\rho}{a}.\frac{a}{r} = \frac{\rho}{a}.\sin D, \qquad \text{or} \qquad \pi = \frac{\rho}{a}.D$$

The corrections listed by Mayer (in Table 8) are the differences

$$D - \pi = D - \frac{\rho}{a}D = \left(1 - \frac{\rho}{a}\right)D$$

They correspond to the two particular cases $D=54'$ and $D=60'$. For example, if $\phi=45°$ and $(a-b)/a=1/200$, it is easily deduced from the first of the two equations cited above, that:

$$\frac{\rho}{a}=\frac{1-e^2}{1-\frac{1}{2}e^2}=0{\cdot}9975$$

hence the amounts which require to be subtracted from D to obtain π are $0{\cdot}0025\times54\times60''=8{\cdot}1''$ and $0{\cdot}0025\times60\times60''=9{\cdot}0''$ for $D=54'$ and $60'$ respectively, as Mayer found.

The three equations cited by him in connection with his fourth precept (p. 76) will be unfamiliar to the modern reader since they are associated with the now out-dated Nonagesimal Method of determining the circumstances of a solar eclipse. De La Caille,[117] and later de Lalande,[118] have discussed this method in some detail, but since their respective explanations are rather obscure—at any rate, to a modern reader—the following discussion has been developed independently in order to exhibit more clearly the reasoning which lies behind the derivation of the equations in question. The parallaxes in longitude and latitude, cited in precept five (p. 76) are deduced from the same basis, as will be demonstrated too.

The celestial horizon, equator and ecliptic are all represented in Fig. 10, in which a broken line is used to indicate great circle arcs lying to the east of the observer's meridian PZ through the poles P, Z of the equator and horizon respectively. O is the point of intersection of the horizon and ecliptic on the west side of that meridian, and ♈, ☊ are the Vernal

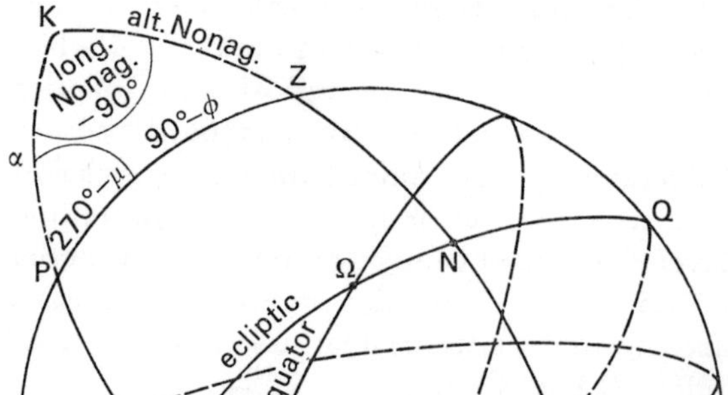

Fig. 10

and Autumnal Equinoxes, i.e. the points of intersection of the equator and ecliptic. In this particular illustration, ♈ is on the east and ♎ on the west of the meridian. N is the nonagesimal—that is, the point on the ecliptic whose celestial longitude is 90° greater than that of O—and so the vertical circle NZ passes through K, the pole of the ecliptic, which is actually situated to the east of the meridian but drawn upwards above the celestial sphere for the purpose of obtaining a clearer illustration. If B is the point at which this great circle arc cuts the horizon to the west of the meridian, it therefore follows that the altitude of the Nonagesimal point $NB = ZB - ZN = ZB - (KN - KZ) = KZ$, since $BZ = KN = 90°$; its longitude measured eastwards along the ecliptic is ♈N.

Let A be a point on the ecliptic 90° east of ♈; then ♈A = 90° and the hour circle AP passes through K. Q is the culminating point of the ecliptic, which lies on the observer's meridian PZ: its right ascension μ = ♈Q = ♈A + AQ = $90° + \widehat{APQ}$. The problem now is: to express ♈N and NB = KZ in terms of this angle μ (which can easily be found if one knows the time and the right ascension of the sun), the corrected latitude ϕ, and the obliquity of the ecliptic α.

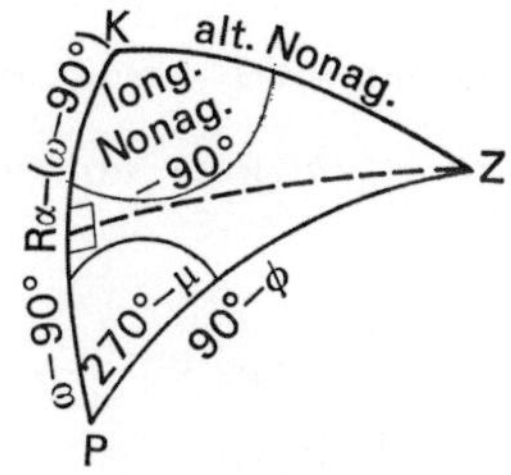

Fig. 11

In the spherical triangle PKZ (Fig. 11), $PZ = 90° - \phi$; $PK = \alpha$, the obliquity of the ecliptic

$$\widehat{KPZ} = 180° - \widehat{APZ} = 180° - \widehat{APQ} = +180° - (\mu - 90°) = 270° - \mu$$

$$\widehat{PKZ} = AN = \text{♈}N - \text{♈}A = \text{long. nonag.} - 90°$$

The solution for the celestial longitude and the altitude of the nonagesimal point is most easily effected by dividing the triangle into two right-angled triangles by means of the perpendicular ZR from Z to the great circle arc PK. If we put $AR = \omega$, $RP = AR - AP = \omega - 90°$, hence $RK = PK - PR = \alpha - (\omega - 90°) = 90° - (\omega - \alpha)$, and the formulae which appear in Mayer's treatise can now be derived as follows:

In the right-angled △PRZ one finds using Napier's rule for circular parts (cf. Fig. 12): viz. sine (middle part) = product of tangents of adjacent

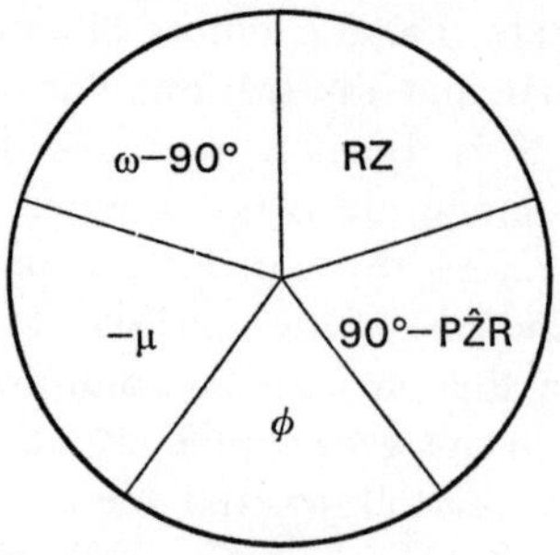

Fig. 12

parts, that

$$\sin(-\mu) = \tan(\omega - 90°)\tan\phi$$

$$-\sin\mu = -\cot\omega\tan\phi,$$

or

$$\tan\omega = \frac{\tan\phi}{\sin\mu}$$

This defines the angle ω in terms of the known angles ϕ and μ. Applying the same rule to the right-angled $\triangle$KRZ (cf. Fig. 13), one obtains:

$$\sin(90 - \overline{\omega - \alpha}) = \tan(180° - \text{long. nonag.})\tan \text{RZ}$$

$$\cos(\omega - \alpha) = -\tan(\text{long. nonag.})\tan \text{RZ}.$$

But from $\triangle$PRZ,

$$\sin(\omega - 90°) = \tan(-\mu)\tan \text{RZ}$$

One finds by eliminating tan RZ:

$$\frac{\cos(\omega - \alpha)}{-\cos\omega} = \frac{-\tan(\text{long. nonag.})}{\tan(-\mu)}$$

$$\tan(\text{long. nonag.}) = \frac{\tan(-\mu)\cos(\omega - \alpha)}{\cos\omega} = -\frac{\tan\mu\cos(\omega - \alpha)}{\cos\omega}$$

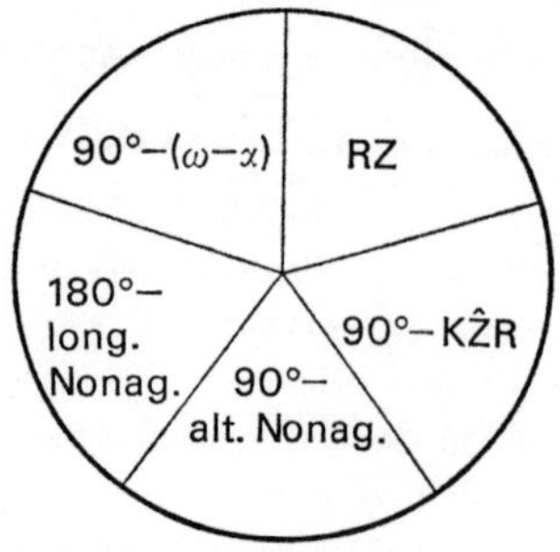

Fig. 13

The omission of the negative sign may have been due to an error in transcription or typography, although one cannot exclude the possibility that it was incurred by Mayer himself.[119]

Similarly, from $\triangle$KRZ:

$$\sin (90° - \text{alt. nonag.}) = \cos (90° - \overline{\omega - \alpha}) \cos \text{RZ}$$

and from $\triangle$PRZ:

$$\sin \phi = \cos (\omega - 90°) \cos \text{RZ}.$$

Eliminating cos RZ:

$$\frac{\cos (\text{alt. nonag.})}{\sin \phi} = \frac{\sin (\omega - \alpha)}{\sin \omega}$$

or

$$\cos (\text{alt. nonag.}) = \frac{\sin \phi \sin (\omega - \alpha)}{\sin \omega}$$

This is the formula given in the text, except for the fact that—owing to an error on the part of either Lichtenberg or Dieterich—the undefined symbol π appears instead of ϕ, the same mistake as was incurred in the above expression for tan ω (see iv, p. 25).

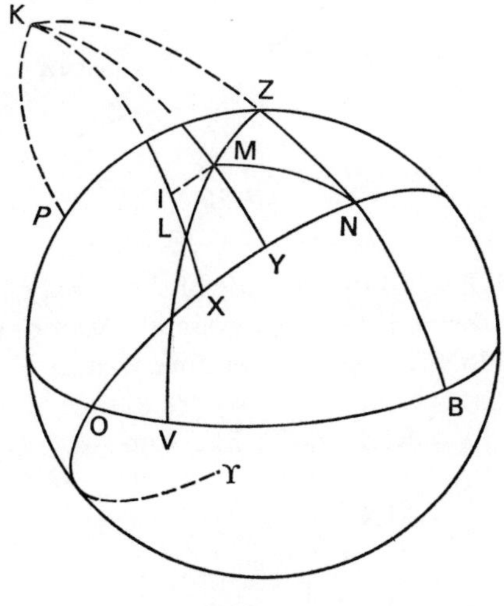

Fig. 14

These formulae for the longitude and altitude of the nonagesimal, which implicitly contain the corrections for astronomical refraction, parallax in terrestrial latitude, and the spheroidal shape of the Earth, are sufficient to uniquely specify the position of N on the celestial sphere. If, in Fig. 14, M denotes the similarly reduced 'true' position of the

Moon, whose geocentric ecliptic co-ordinates for any given time can be interpolated from lunar ephemerides, the angular distance NM is known and the next stage in the procedure is to find the effect of the parallax upon the celestial longitude and latitude of M and thereby obtain the values of the apparent ecliptic co-ordinates required for the graphical construction of the solar eclipse. It will be evident from our previous considerations that the apparent position of the Moon L lies on the same vertical circle ZV as M does, and that the effect of parallax in latitude is to decrease the zenith distance of the moon from ZL to ZM, the difference being $ML = p \sin ZL$. Let KLX, KMY be the latitude circles drawn from the pole of the ecliptic K through L, M to meet the ecliptic ON at X, Y respectively, as indicated in Fig. 14. MI is the small circle arc through M, parallel to the ecliptic, meeting KLX at I. Then XY is the parallax in the Moon's longitude, and IL that in its latitude, which must now be evaluated in terms of known quantities.

Consider the spherical triangles KMZ and ZMN which together form the quadrantal triangle KMN (Fig. 15). KM is the known co-latitude of M; MN the known distance between the true positions of the Moon and

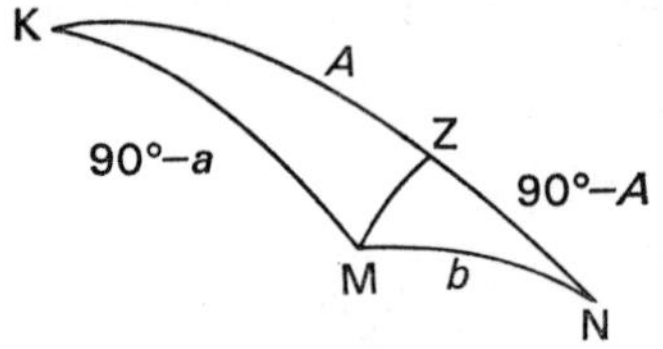

Fig. 15

the nonagesimal; KZ = NB the altitude of the nonagesimal; and KN = 90° the latitude circle through the nonagesimal. Adopting the notation used by Mayer in his fifth precept, we therefore write:

$$KM = 90° - MY = 90 - a$$

$$MN = b$$

$$KZ = A$$

and so

$$ZN = KN - KZ = 90° - A$$

It will now be demonstrated that the required parallaxes in longitude and latitude at the time of a solar eclipse (when the Moon's celestial latitude is effectively zero) can be expressed solely in terms of these known quantites and the Moon's horizontal parallax π.

$$\text{Parallax in longitude XY} = \frac{\text{IM}}{\cos \text{MY}} = \frac{\text{ML} \cos \widehat{\text{IML}}}{\cos a}$$

$$= \frac{p \sin \text{ZL}}{\cos a} \sin \widehat{\text{LMY}} = \frac{p \sin(\text{ZM}+\text{ML})}{\cos a} \sin \widehat{\text{KMZ}}$$

$$= \frac{p}{\cos a} \sin \text{ZM} \cos \text{ML} \sin \widehat{\text{KMZ}}$$

$$+ \frac{p}{\cos a} \cos \text{ZM} \sin \text{ML} \sin \widehat{\text{KMZ}}$$

Since ML is small, to a sufficient approximation we may write

$$\cos \text{ML} = 1$$

and

$$\sin \text{ML} = \text{ML} = \text{ML} \sin 1'' = p \sin \text{ZL} \sin 1'' = sp \sin \text{ZL}$$

where $s = \sin 1'' = 1/206265$ radian and p is expressed in seconds of arc. Hence

$$\text{XY} = \frac{p}{\cos a} \sin \text{ZM} \sin \widehat{\text{KMZ}} + \frac{sp^2}{\cos a} \sin \text{ZL} \cos \text{ZM} \sin \widehat{\text{KMZ}}$$

In $\triangle$KLZ,

$$\frac{\sin \text{ZM}}{\sin \hat{\text{K}}} = \frac{\sin \text{KZ}}{\sin \widehat{\text{KMZ}}}$$

$$\sin \text{ZM} \sin \widehat{\text{KMZ}} = \sin A \sin \hat{\text{K}} \quad \text{or} \quad \sin \widehat{\text{KMZ}} = \frac{\sin A}{\sin \text{ZM}} \sin \hat{\text{K}}$$

In $\triangle$NMZ,

$$\cos \text{ZM} = \cos \text{MN} \cos \text{ZN} + \sin \text{MN} \sin \text{ZN} \cos \hat{\text{N}}$$

$$= \cos b \sin A + \sin b \cos A \cos \hat{\text{N}}$$

In $\triangle$KMN,

$$\cos \text{KM} = \cos \text{KN} \cos \text{MN} + \sin \text{KN} \sin \text{MN} \cos \hat{\text{N}}$$

$$\cos (90 - a) = \sin b \cos \hat{\text{N}} \quad \text{or} \quad \cos \hat{\text{N}} = \frac{\sin a}{\sin b}$$

$$\frac{\sin \hat{\text{K}}}{\sin \text{MN}} = \frac{\sin \widehat{\text{KMN}}}{\sin 90^\circ} \quad \text{or} \quad \sin \hat{\text{K}} = \sin b \sin \widehat{\text{KMN}}$$

From these formulae, it is easy to deduce that:

$$\sin \widehat{\text{KMZ}} = \frac{\sin A \sin b \sin \widehat{\text{KMN}}}{\sin \text{ZM}}$$

$$\cos \text{ZM} = \cos b \sin A + \sin a \cos A$$

Substituting in the equation for XY yields:

$$\mathrm{XY}=\frac{p}{\cos a}\sin A\sin b\sin \widehat{\mathrm{KMN}}+\frac{sp^2}{\cos a}\frac{\sin \mathrm{ZL}}{\sin \mathrm{ZM}}$$

$$\times(\cos b\sin A+\sin a\cos A)\sin A\sin b\sin \widehat{\mathrm{KMN}}$$

Now

$$\frac{\sin \mathrm{ZM}}{\sin \mathrm{ZL}}=\frac{\sin(\mathrm{ZL}-\mathrm{ML})}{\sin \mathrm{ZL}}=\frac{\sin \mathrm{ZL}\cos \mathrm{ML}-\cos \mathrm{ZL}\sin \mathrm{ML}}{\sin \mathrm{ZL}}$$

and since

$$\mathrm{ML}=p\sin \mathrm{ZL}\leqslant p\simeq 0{\cdot}017$$

this approximates to:

$$1-\frac{\mathrm{ML}\cos \mathrm{ZL}}{\sin \mathrm{ZL}}=1-\frac{\mathrm{ML}}{\tan \mathrm{ZL}}\geqslant 1-\mathrm{ML}=0{\cdot}983$$

In general, therefore, it is justifiable to equate the ratio sin ZL/sin ZM to unity; and in particular, when the Moon is on the horizon,

$$\frac{\sin \mathrm{ZL}}{\sin \mathrm{ZM}}=1 \ \textit{exactly}$$

To a sufficient approximation, therefore, p may be replaced by π, the Moon's horizontal semi-diameter; hence

$$\mathrm{XY}=\frac{\sin \mathrm{KMN}}{\cos a}[\pi\sin A\sin b+s\pi^2\sin^2 A\sin b\cos b$$

$$+s\pi^2\sin A\cos A\sin a\sin b]$$

In the case of a solar eclipse, which is presently under consideration, one must also take account of the fact that the Moon at that time is very close to one of the nodes of its orbit. Mathematically expressed, this condition corresponds to: $a=0$, KL$=90°$ and $\widehat{\mathrm{KLN}}=90°$; also $\sin\widehat{\mathrm{KMN}}=\sin(\widehat{\mathrm{KLN}}-\widehat{\mathrm{LKM}})=\sin\widehat{\mathrm{KLN}}\cos\widehat{\mathrm{LKM}}-\cos\widehat{\mathrm{KLN}}\sin\widehat{\mathrm{LKM}}$ and since $\widehat{\mathrm{LKM}}=\mathrm{XY}$ is a very small quantity, this identity reduces to the approximate form:

$$\sin\widehat{\mathrm{KMN}}=\sin\widehat{\mathrm{KLN}}-\mathrm{XY}\cos\widehat{\mathrm{KLN}}=\sin 90°-\mathrm{XY}\cos 90°=1$$

Introducing these further simplications, one finally obtains:

$$\mathrm{XY}=\pi\sin A\sin b+s\pi^2\sin^2 A\sin b\cos b$$

This is precisely the expression for the parallax in longitude quoted by Mayer.[120] The above analysis has, however, shown that it is only approximately valid, and that the approximation is most reliable when the eclipse is observed shortly after sunrise or shortly before sunset. This same proviso also holds for the corresponding formula for the parallax in latitude, since it is derived with the aid of the equation for XY, as will now be shown.

$$\text{Parallax in latitude IL} = \text{IM} \cot \widehat{\text{ILM}} = \frac{\text{XY}}{\cos \text{MY}} \cot \widehat{\text{KMZ}}$$

$$= \frac{\text{XY}}{\cos a} \frac{\cos \widehat{\text{KMZ}}}{\sin \widehat{\text{KMZ}}}$$

In $\triangle$KMZ,

$$\cos \text{KZ} = \cos \text{KM} \cos \text{ZM} + \sin \text{KM} \sin \text{ZM} \cos \widehat{\text{KMZ}}$$

$$\cos \widehat{\text{KMZ}} = \frac{\cos A - \sin a \cos \text{ZM}}{\sin a \sin \text{ZM}}$$

Hence

$$\frac{\cos \widehat{\text{KMZ}}}{\sin \widehat{\text{KMZ}}} = \frac{\cos A - \sin a \cos \text{ZM}}{\cos a \sin \text{ZM}} \times \frac{\sin \text{ZM}}{\sin A \sin \hat{\text{K}}}$$

Using the expressions previously derived for cos ZM and sin $\hat{\text{K}}$, we get

$$\text{IL} = \frac{\text{XY}}{\cos a} \left[\frac{\cos A - \sin a (\cos b \sin A + \sin a \cos A)}{\cos a \sin A (\sin b \sin \widehat{\text{KMN}})} \right]$$

$$= \frac{\text{XY}}{\sin \widehat{\text{KMN}}} \left[\frac{\cos A (1 - \sin^2 a) - \sin a \cos b \sin A}{\cos^2 a \sin A \sin b} \right]$$

$$= \frac{\text{XY}}{\sin \widehat{\text{KMN}}} \left[\frac{\cos A}{\sin A \sin b} - \frac{\sin a \cos b}{\cos^2 a \sin b} \right]$$

$$= \frac{1}{\cos a} [\pi \sin A \sin b + s\pi^2 \sin^2 A \sin b \cos b$$

$$+ s\pi^2 \sin A \cos A \sin a \sin b] \times \left[\frac{\cos A}{\sin A \sin b} - \frac{\sin a \cos b}{\cos^2 a \sin b} \right]$$

$$= \pi \left[\frac{\cos A}{\cos a} - \frac{\sin A \sin a \cos b}{\cos^3 a} \right] + s\pi^2 \left[\frac{\sin A \cos b \cos A}{\cos a} - \frac{\sin^2 A \sin a \cos^2 b}{\cos^3 a} \right]$$

$$+ s\pi^2 \left[\frac{\cos^2 A \sin a}{\cos a} - \frac{\sin A \cos A \sin^2 a \cos b}{\cos^3 a} \right]$$

When $a=0$, $\sin a=0$, $\cos a=1$; so we finally obtain the required result for the parallax in latitude, namely:

$$\pi \cos A + s\pi^2 \sin A \cos A \cos b$$

There is another term in the corresponding equation used by Mayer, but this contains $\sin a$ as a factor and therefore should vanish for the particular case of a solar eclipse. Since $A \leqslant 90°$, and $b \leqslant 180°$, the term in π^2 can take on a negative value only when b

is obtuse. This case has been taken into account by Mayer in writing down the formula for the parallax in latitude; but for the sake of consistency, it ought to be either allowed for in that for the parallax in longitude as well, or omitted altogether. This latter alternative is preferable since in the application of these two formulae we should be concerned not merely with the magnitude of the circular functions of the angular distance b, but also with their respective signs. As Mayer points out, the finite distance of the Sun can be taken into account by subtracting the Sun's horizontal parallax from that of the Moon.

The Moon's apparent longitude ♈X and latitude LX are now given by:

♈X = ♈Y − XY = Moon's true longitude − parallax in longitude

LX = IX − IL = MY − IL = Moon's true latitude − parallax in latitude

for the figure which we have used, in which the Moon is to the west of the nonagesimal: the parallaxes would both require to have been added had the Moon been situated to the east of that point.

Having calculated the Moon's apparent position ($\lambda_{☾}$, $\beta_{☾}$) at three different instants close to the time of the eclipse, and the Sun's longitude ($\lambda_{\odot}$) for these same instants from his solar tables, Mayer was able to interpolate and tabulate the differences ($\Delta\lambda = \lambda_{\odot} - \lambda_{☾}$) between the apparent longitudes of the Sun and Moon, and the values of the apparent latitudes of the Moon, for five-minute intervals throughout the course of the eclipse. Alongside these he wrote down the corresponding figures for the sum of the semi-diameters of the Sun and Moon, which he might either have obtained from observations made close to the time of the eclipse or from tables. This sum, which he denotes by c, is corrected for augmentation by means of the formula derived above (p. 28): very probably it also contains empirical corrections for the observed effects of irradiation in increasing the Moon's and Sun's apparent semi-diameters, which de Lalande later estimated as 2″ and $3\frac{1}{2}$″.[121]

The times of the beginning and end of the eclipse are obtained simply by finding the instants at which the condition $\sqrt{c^2 - x^2} = \Delta\lambda$ is satisfied (cf. Fig. 16), x being the value of the Moon's latitude at those instants.

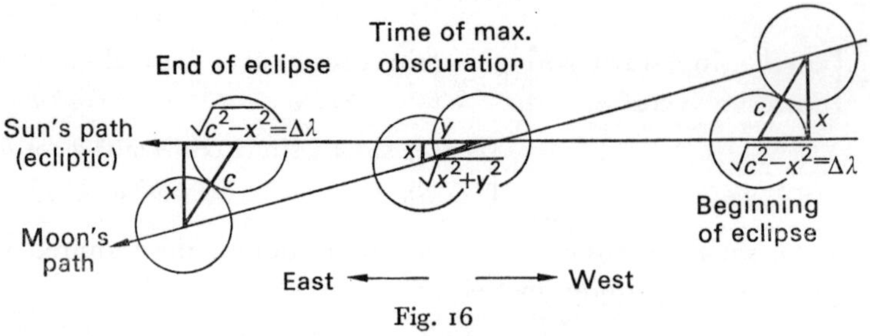

Fig. 16

This equation can be solved by a method of successive approximations, or by supposing that an error (= dy, say) in $\Delta\lambda$ is caused by an error dx in the value initially assumed for x. The connection between dy and dx

is then found from the condition (cf. Fig. 17):

$$c^2 = (x + dx)^2 + (\sqrt{c^2 - x^2} - dy)^2$$

If one neglects second order quantities in dx and dy, this reduces to $\sqrt{c^2 - x^2}/x = dx/dy$, the very relationship quoted by Mayer in his seventh precept with the aid of which the acquired instants are easily determined.

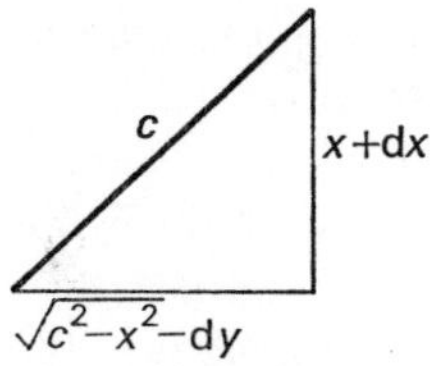

Fig. 17

The time of maximum obscuration is obtained by a similar procedure, as described in the text; the maximum quantity of the eclipse being defined by Mayer as $6(c - \sqrt{x^2 + y^2})$/semi-diameter of Sun. A worked example illustrates this latter part of the procedure, but is perfectly straightforward and consequently requires no commentary.

IV. *Treatise on the relationship of colours (1758)*

This is the treatise to which Lichtenberg devotes his longest commentary —perhaps because it seemed to him to constitute a test case for his view (at that time) that reality can be attained only through mathematics, by excluding the distorting influence of the senses.[122] Inasmuch as he believed clear and distinct ideas to be intimately related to our concepts of ratios and proportions, Lichtenberg was adopting the Cartesian approach to knowledge.[123] His avowed acceptance of Newtonian physics seems to have been governed by the fact that it possessed a high degree of certainty by virtue of its being grounded upon mathematics; but he was later to recognize that Newton's discoveries were due just as much to an essentially non-mathematical leap of induction, and this made him become sceptical of the power of mathematical deduction.[124] Although Mayer questioned the Newtonian interpretation of the prismatic spectrum on the basis of his own experience in working with colours,[125] he also subjected the phenomenon of colour to exact measurement; and it was surely this which attracted Lichtenberg to the subject, and caused him to examine the empirical basis on which Mayer's theory of colour-mixing had been founded. His views are clearly expressed in his commentary, and need not be repeated here.

A circumstance which Lichtenberg fails to mention, however, is that there exist among Mayer's papers three German manuscripts which are closely concerned with this question. Two of these, entitled *Mathematische Untersuchungen der Farbenmischungen*[126] and *Von den Farben*[127] appear to be incomplete drafts of the Latin treatise presently under discussion; but the third, *Bemerkungen und Excerpte über Farben*,[128] though also incomplete and unsystematically arranged, is of some interest since most of it was written on 22 June 1759 (over seven months after the date of the open lecture at which this Latin treatise was presented). The *Bemerkungen* clearly reveal that, as a result of subsequent experimentation, Mayer had by then come to appreciate the necessity of introducing further factors into his theory; namely, the strengths of colouring ('Kraft zu färben' or *vires tingendi*) possessed by different colours, and the respective weights of the colour pigments.

One can therefore not discount the possibility that it was his reading of this manuscript and not his familiarity with Lambert's colour-pyramid which initially inspired Lichtenberg to generalize Mayer's original theory by weighting the proportions of the primary colours in the inverse ratio of their strengths of colouring and in direct proportion to their masses.[129]

Mayer's final remarks, dated 20 October 1759, are concerned with the problem of shading. By now, he had come to regard all kinds of shadow colours—obtained 'mostly from used or unused umbra, mountain-blue, rather less cinnabar and pine soot, etc.'—as mixtures of orange-yellow and the primary colour blue, which he denotes as $r^6g^6 = f$ and b^{12} respectively.[130] The mixture $r^2g^2b^8$ is referred to as 'complete black', and $r^4g^4b^4$ as 'the whole shadow-colour.' On the assumption that all shadow-colours possess the same strength of colouring, they are mixed with the local colour in such a way as to produce 'the right degree of shading for this local colour.' The proportions chosen by Mayer, although he describes them as 'arbitrary', were presumably based upon the results of repeated experimentation. From two pages of notes bound at the end of *Von den Farben* although they appear to be more closely concerned with the problem of shading, it is evident that Mayer was also well aware of the influence of the background and distance of an object upon the strength and quality of the shadows which it casts. The practical difficulties involved in attempting to reduce to strictly mathematical terms factors such as these, and the colouring properties of so many different chemical mixtures—recipes for which are contained in the *Excerpte* . . .—effectively counteract the advantages of Mayer's theory of colour-mixing as a basis for the naturalistic representation of an object or work of art. On the other hand, it does provide a useful guide to the formulation of empirical rules that might have been applied by different artists to produce a greater degree of consistency in their artistic reproductions. Yet, despite Mayer's efforts to encourage others to take up this form of art, there is no evidence to suggest that it ever came into widespread use. The fundamental question of whether colour can ever be completely described in purely mathematical terms, which Isaac Newton had believed to be the case[131] and against which Johann Wolfgang von Goethe was later to rebel,[132] has been answered in the negative only during the present century through Werner Heisenberg's discovery of the Uncertainty Principle; for this has clearly revealed that it is impossible *in principle*, as well as in practice, to separate the subjective and objective aspects of reality in any observation of the physical world.[133]

V. *Treatise on the new catalogue of zodiacal stars (1759)*

The main feature of interest in this treatise is the method used by Mayer for finding the right ascensions of the stars included in his catalogue, which he describes in the text as 'the most reliable, and the one best suited both to the observations and to the instruments'. Nevertheless, if one examines it carefully, one finds that its merits are not to be rated as highly as this statement would appear to suggest. Essentially, Mayer deduces the value of the right ascension of a star from a knowledge of its difference in right ascension from that of the Sun, and of the right ascension of the Sun itself. The advantage of this difference method is that it does not suffer from the fact that the plane of the mural quadrant may not coincide exactly with that of the meridian. Its disadvantage, however, is that the direction of the Sun's diurnal motion is supposed to be parallel to that of the star which, on account of the ecliptic being inclined at an angle of about 23° 27′ to the equator, is not strictly true. Mayer is, of course, justified in supposing that no appreciable error will arise on this account when the Sun approaches either of the solstices, where it attains its maximum and minimum declinations of +23° 27′ and −23° 27′ respectively; the Sun's right ascension near these times being computed

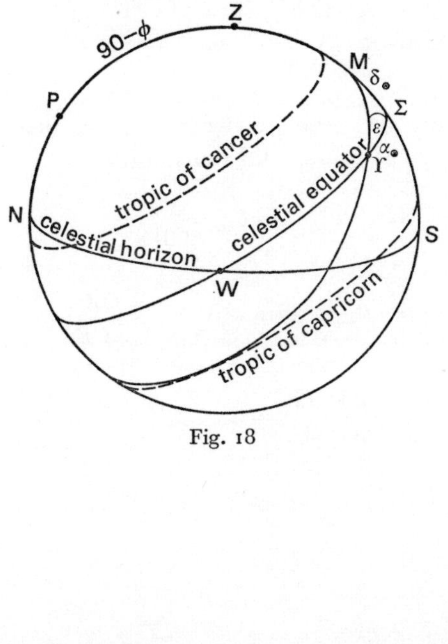

Fig. 18

from solar tables whose elements had been corrected with the aid of transit observations made close to one of the equinoxes.

The figure on the previous page (Fig. 18), which corresponds to an observation made at local apparent noon by an observer in north latitude ϕ on a day sometime in May before the Sun reaches the tropic of cancer, should help to clarify the principle involved in the calculation of the Sun's right ascension. ♈ denotes the Vernal Equinox, and M the position of the Sun as it crosses the observer's meridian. For convenience, let us put the right ascension of the Sun $\Upsilon\Sigma = \alpha_\odot$, its declination $M\Sigma = \delta_\odot$, and the obliquity of the ecliptic $\widehat{M\Upsilon\Sigma} = \epsilon$.

Now, in the text of the treatise under discussion, Mayer makes a statement to the effect that $\alpha_\odot$ was found by applying the [meridian] zenith distance. The conventional means of doing this would be first, to find $\delta_\odot$ from the observed zenith distance of the sun at the time of its meridian transit ($=\mathcal{Z}_\odot$ say), using the relation:

$$\delta_\odot = M\Sigma = 90° - PZ - ZM = 90° - (90° - \phi) - \mathcal{Z}_\odot = \phi - \mathcal{Z}_\odot$$

and then, to substitute it into the equation

$$\sin \alpha_\odot = \frac{\tan \delta_\odot}{\tan \epsilon}$$

(found from a consideration of the $\triangle \Upsilon M\Sigma$), to obtain $\alpha_\odot$.

Mayer is much more likely to have found $\alpha_\odot$ from $\delta_\odot$ and ϵ, however, by employing the graphical method described in his lecture notes,[134] which is essentially as follows (Fig. 19):

(i) Draw a circle, centre C, and its conjugate diameters $\Upsilon\Omega$ and ΣW.

(ii) Construct $\widehat{C\Sigma B} = 90° - \epsilon$, and produce ΣB to meet $\Omega\Upsilon$ produced in B.

(iii) Construct $\widehat{CBG} = \delta_\odot$, and produce BG to meet $W\Sigma$ in G.

(iv) Through G, draw $aGA \parallel \Omega C\Upsilon$ to intersect the circle at a, A.

(v) $\widehat{\Upsilon A}$ or $\widehat{\Upsilon a} = \alpha_\odot$.

It is easy to supply the proo of the trigonometrical relationship cited above which Mayer, of course, did not require; for in Fig. 19,

$$\tan \delta_\odot = \frac{GC}{CB}$$

and

$$\tan \epsilon = \cot (90° - \epsilon) = \frac{\Sigma C}{CB};$$

so

$$\frac{\tan \delta_\odot}{\tan \epsilon} = \frac{GC}{CB} \times \frac{CB}{\Sigma C} = \frac{GC}{\Sigma C} = \frac{GC}{CA} = \sin \widehat{GAC} = \sin \widehat{AC\Upsilon} = \sin \widehat{\Upsilon A} = \sin \alpha_\odot$$

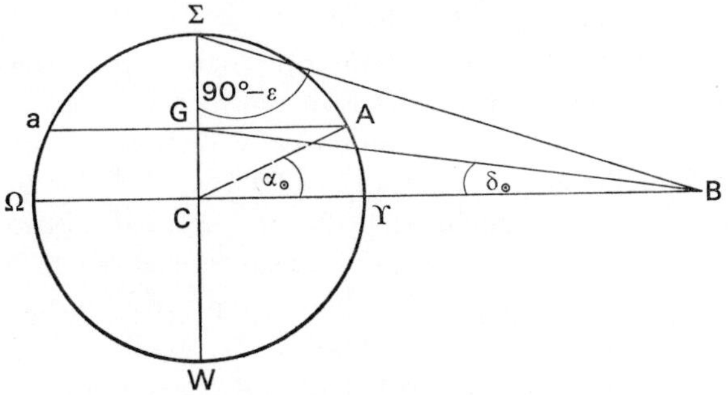

Fig. 19

Mayer points out that when $\delta_{\odot}$ is northerly, G is above C (in Fig. 19); but when $\delta_{\odot}$ is southerly, G is below C. He also explains that $\widehat{\Upsilon A} = \alpha_{\odot}$ when $\delta_{\odot}$ is increasing (i.e. between the winter and summer solstices), whereas $\widehat{\Upsilon a} = \alpha_{\odot}$ when $\delta_{\odot}$ is decreasing.

Whichever way the result is obtained, the essential point to note is that a knowledge of $\alpha_{\odot}$ necessarily presumes that of ϕ and ϵ. This is not the impression which one gains from a superficial reading of Mayer's treatise, in which the wording is such as to suggest that an accurate knowledge of ϕ is required only for the determination of the stellar declinations, and that only the ecliptic co-ordinates of the stars—and not the equatorial ones—will be affected by the known secular variation of ϵ (estimated by Mayer as 44″ per century).

An accurate reduction of Mayer's catalogue of zodiacal stars to *ecliptic* co-ordinates was made some 25 years after his death—12 years after the appearance of the *Opera inedita* . . .—by J. A. Koch,[135] who estimated an accuracy of within 10″ for the star positions which he had reduced to the beginning of the year 1778 (being the mean epoch between 1756 and 1800), despite the fact that he had adopted the same value of $\epsilon = 23°\,28'\,16''$ found by Mayer from his own observations and had deliberately made no allowance for the still-uncertain secular variation in the celestial longitudes of the stars. The first complete catalogue of all the astronomical observations made by Mayer between 1756 and 1761 (inclusive) was published by order of the Commissioners of Longitude in 1826:[136] its contents were critically discussed a few years later by Francis Baily in a paper read to the Astronomical Society of London on 11 June 1830.[137] A revised edition, in German, appeared during the last decade of the nineteenth century.[138]

VI. *Treatise on the proper motion of the fixed stars* (*1760*)

This treatise is the first known attempt to seek a quantitative explanation of the changes in star positions detected by Edmond Halley in 1718.[139]

Historically, its importance lies in the fact that it provided William Herschel with the observational data from which he deduced that the solar system is moving through space in the direction of the star λ Herculis,[140] a conclusion which is diametrically opposed to that stated by Mayer himself in the text. The reason why Herschel was able to detect a systematic drift in the stellar co-ordinates whereas Mayer did not, is due to his adoption of the hypothesis (which turned out to be false) that brighter stars are nearer to the Earth (or Sun) and fainter ones farther away. Since the *nearer* stars had to exhibit the greatest amount of movement due to this cause, Herschel based his analysis upon the values of the proper motions of the twelve *brightest* stars in Mayer's treatise, which he found conveniently tabulated in the second edition of Lalande's *Astronomie*.[141] Although the reality of the solar motion through space could thereafter no longer be doubted, its cause was still a matter for speculation. Lalande himself had already expressed his belief that such a phenomenon might arise from a tangential, as well as a radial, force being associated with the Sun's axial rotation,[142] but Pierre Prévost preferred to regard it as a consequence of the difference in the net gravitational attraction between one hemispherical region of space and the opposite one.[143] Theoretical speculations on the cause of this motion were reviewed by Prévost at about the same time,[144] and Francis Baily drew attention to the observational problem in a memoir on the subject written some fifty years later.[145]

In principle, the validity of Prévost's hypothesis can be tested by determining whether the solar motion is accelerative and curvilinear, as is required; in practice, however, the accuracy of the early observations on which Mayer's figures for the proper motion had necessarily been based, was not adequate for this purpose. Somewhat optimistically, Prévost and Maurice made a similar comparison of the right ascensions and declinations of stars observed by Mayer with those deduced from observations made in 1797, in the hope of obtaining more reliable results; but despite the great care which they took in eliminating instrumental errors and in calculating variations in proper motion, precession, etc. they were unable to attain the degree of accuracy necessary for testing their supposition that the solar system is travelling in a conic section.[146] Among the other complicating factors which enter into this analysis is the difficulty of estimating stellar distances and the fact that each star has a peculiar motion of its own besides that which it appears to have as a result of the solar motion. The statistical procedures which are normally used in the investigation of stellar motions introduce further assumptions and uncertainties which, together with the revolution in our concepts of the nature of space and time that has resulted from the discovery of non-euclidean geometries and from Einstein's theory of relativity, are sufficient to preclude the possibility of the true cause of these movements becoming known in the foreseeable future, as Mayer has correctly surmised.

The *Opera inedita* made a favourable impression on George III, to whom it was dedicated, when he was presented with one of the earliest printed copies of it by Lichtenberg on 26 October 1774. He praised Dieterich's printing, which he thought was excellent and worthy of Mayer's writings. After communicating this fact to his friend Dieterich four days later, Lichtenberg goes on to give his own opinion of the work, remarking: 'I was delighted with the book, except in this respect, that the lunar map and the triangle [Tab. III = Plate 5] have been broken up by the bookbinder. This is an unfortunate oversight and must certainly be remedied, if you are having any more bound. I told you, as you may remember, that a format of that particular size was chosen largely on account of the lunar map, and if it is broken up, my whole purpose will be defeated, as you can see. Be so kind as to send copies—well chosen ones, moreover—to Lambert and to Bernoulli in Berlin, Röhl at Greifswald, Ljungberg at Kiel, Father Hell in Vienna, De la Lande in Paris, Niebuhr, my brothers, and to Mayer's widow. With every copy write to say that it was at my request'.[147]

Only a few weeks later, Lichtenberg was honoured by an invitation from the king to stay with the Royal Family at Kew, during which time he often visited the observatory at Richmond. He returned to Lord Boston's house in the West End of London in February 1775, and continued to mix freely with astronomers, politicians, and aristocrats. On 2 March of the same year he was introduced by the Astronomer Royal Nevil Maskelyne and the moral and political philosopher Dr. Richard Price, to a meeting of the Royal Society; and spent the whole of the following Sunday with Maskelyne at the Greenwich Observatory.[148] In the course of the summer months, he made a series of journeys to other parts of England including Oxford, where he became acquainted with the first Radcliffe Observer Dr Thomas Hornsby whose observatory, in his opinion, surpassed the one in Richmond by as much as the latter surpassed that in Göttingen.[149] Hornsby too apparently thought very highly of Mayer's work and of the *Opera inedita* in particular.[150] Further information regarding Lichtenberg's activities during his 15-month stay in England can be found in a special study by Margaret L. Mare and W. H. Quarrell of *Lichtenberg's Visits to England as described in his Letters and Diaries* (Oxford, 1938). According to these literary scholars, the climax of his second visit was his experience of the English stage; but his personal acquaintance with the Genovese geologist Jean André de Luc and the English chemist Joseph Priestley during this time was also to have a strong influence upon the meteorological and physiological views which he subsequently adopted.

Lichtenberg, who by now had been raised to the status of an Ordinary Professor at Göttingen, returned there on 31 December 1775, bringing with him three new English pupils. After that time, he never left Göttingen again except for two short journeys to Hamburg and to Gotha. For the next year most of his working hours were devoted to lecture preparation. Then in 1777, two circumstances combined to produce a marked change in his scientific interests, putting paid to any thoughts which he may still have entertained about editing the proposed second volume of Mayer's *Opera inedita*; namely, his discovery of the electrical figures which still bear his name,[151] and the death (in August on that year) of his friend and university colleague Johann Christian Polycarp Erxleben.[152] His consequent interest in the science of electricity and increasing involvement in the course of lectures in experimental physics which Erxleben himself had previously been giving,[153] made Lichtenberg less inclined to pursue astronomical and mathematical topics of research. He used Erxleben's *Anfangsgründe der Naturlehre* (Göttingen, 1772) as the basis of his lectures, editing and republishing this text book on four occasions (in 1784, 1787, 1791 and 1794) and each time enlarging it with new notes and experimental results of his own. Although Lichtenberg never realized his intention of composing an entirely new text book of his own, one of his students, Gottlieb Gamauf, took down a whole course of his lectures more or less verbatim, and had them published nine years after his teacher's death.[154] They reveal that Lichtenberg was very conscious in his teaching, of the methodological aspect of science—or more precisely, undogmatic enquiry into all phenomena, and the constant exercise of one's critical faculty—and continually questioned not only hypotheses themselves, but one's ability to hypothesize. His lectures and demonstrations, which at times attracted over a hundred students, brought him such great fame that George III decided to send his three youngest sons to Göttingen where they finished their education as Lichtenberg's private pupils in physics. This had not a little to do with the fact that he received the title of *Hofrath* soon afterwards (in 1788).

A further reason for Lichtenberg's failure to edit the still unpublished treatises of Mayer to which he refers in the *Preface* to the *Opera inedita* was his simultaneous involvement in literary ventures such as Dieterich's *Göttinger Taschen Calender* and Georg Forster's *Göttingisches Magazin für Wissenschaften und Literatur* in which financial considerations were probably a stronger motive than his desire for self-expression. The numerous articles and notes which he wrote for these periodicals, Heinrich Christian Boie's *Das deutsche Museum* and the *Neues Hannoverisches Magazin* resulted in his acquiring a well-deserved reputation among his contemporaries as a popularizer of science and literary figure in addition to that which he already enjoyed as an astronomer and experimental physicist. Although this two-sided reputation has since been further enhanced by the posthumous publication of an appreciable portion of his vast correspondence, his ten 'Waste-books', and his physical and mathematical writings,[155] it would appear that posterity is more inclined to remember him for his literary achievements than for his contributions to science. Lichtenberg, while overtly agreeing with Mayer's view that it was better to know a little thoroughly than a great deal superficially,[156] lacked himself the strength of purpose to resolutely challenge each scientific problem or human predicament as it presented itself. He was better at asking questions than at supplying answers, and his statements about the nature of science reflect the presence of conflicting—or at least competing—elements in the development of his thought which complicate the task of interpreting it consistently. Perhaps a better appreciation of Lichtenberg's rôle in the history of science will therefore be gained by placing greater emphasis upon the fruits, as opposed to the sources of his achievements.[157] If so, the re-publication of the present work will have served a dual purpose.

OPERA INEDITA

PREFACE

After the interruption of the edition of the early transactions of the Royal Society,[158] the fellows undertook the task of otherwise making their own contributions known to the general public, with the result that last year the commission was most kindly entrusted to me to assemble our Mayer's writings for publication. For that reason not only the manuscript treatises formerly presented to the Royal Society by this distinguished author, but also whatever journals, papers and sketches had been bought from Mayer's heirs by the King's revered council in the lands ruled by the Electorate of Brunswick-Luneberg [i.e. by the Hanoverian Government] for use in our observatory, were handed over to me so that I might record from them any remarks which would shed light upon [Mayer's unpublished] treatises, renowned from now onwards throughout the entire world of learning. I immediately took on the task which I believe is not unwelcome to the more cultured part of Europe, and it appears that I have been allowed to do so at the command, and through the privilege of his most august Majesty: and now I present to the public the first volume, albeit a little later than I had initially hoped, owing to various circumstances which it is of no importance to recount here.[159]

Besides the selenographic table bought from his heirs, this work contains those dissertations presented by that distinguished author to the Royal Society, which are to be found complete. Some [others] were not entirely completed by this man, whose very fertile genius was continually intent on the investigation of new truths, and who at that time harboured little hope that his works would achieve widespread circulation, on account of a disagreement which had arisen between the Royal Society and his own bookseller[160]; others he himself had taken back from the libraries of the Royal Society, and these were found mutilated after his death; these will be contained in the second volume, due to come out next year.[161]

It will be helpful to indicate here that the Theory of the magnet,[162] the commentary on the [orbital] motion of Mars[163] and the description of an Astrolabe[164] invented by the author will grace the second volume. The whole of the Theory of the magnet which its author left, survives intact, but in the actual article on the Theory of Mars nothing remains except a historical introduction and tables: and the description of the Astrolabe may be incomplete but, because the selfsame instrument which was made by the author is in Kästner's possession, it can easily be supplemented.[165]

Only a few words of caution will be necessary here concerning the articles contained in this first volume, since I have added the bulk of those comments which I believed to be necessary, in an appendix. There, the reader will find a description of certain works of Mayer that could not appear in this volume. I have commented liberally on these for the sake of continuity, but I am far from believing that they can meet with the approval of the learned friends of Mayer; indeed, I freely confess that certain of them are not satisfying even to myself.[166]

What the verdict might be concerning the chromatic triangle affixed at the end, and its purpose, will be apparent from what I have said in the preface and explained below. Here I merely advise you that it does not bear visual inspection from close range, but only from a suitable distance where the irregularities that arise as much from mistakes in painting as from flaws in the paper, may no longer be seen.

Mayer's catalogue of the fixed stars as well as the lunar tables which he is responsible for perfecting, will be a most welcome gift to astronomy. Indeed, much work will still be required before it can be more generally used, since the author did not add either the changes [in the right ascension and declination] or [celestial] longitudes and latitudes. I have left this task to those who concentrate their energies on computing ephemerides, who are more skilled than I am in this type of calculation and have more leisure.[167] It was, however, entirely my responsibility to make the printed copy correspond as closely as possible to Mayer's manuscript and I am able to assert that this was done with such care that, with the correction of those four errors which I have indicated at the end of the work,[168] I am confident that the catalogue is consistent in every respect with Mayer's manuscript. For after I had exhausted almost all my patience in comparing it with the original, I handed it over together with the manuscript to a man well skilled in this,[169] promising no mean reward for any mistake he might detect, but nothing more could be found. The same holds true of the catalogue annexed [in the original version] to the dissertation on the proper motion of the fixed stars. Thus astronomers now have at their disposal an extremely accurate copy of Mayer's catalogue, a production of the labour of a number of men, which they will be able to transform into a useful form in a shorter space of time than if all the reductions had had to be by me alone.

I have added nothing about the life and writings of Mayer, both because a published elegy to him by Kästner already exists, to which is annexed a very full catalogue of Mayer's writings.[170] This will appear again soon, inserted in Kästner's works.[171] Another principal reason is that, after the publication of the present work, the bookseller decided that all the works of that author [Mayer] should be published in the same form. To this collection, besides [an account of] his life, is prefaced a picture very closely resembling the original painted by the author himself, etched on copper-plate.[172] Written in Göttingen 28 Aug. 1774.[173]

I. A MORE ACCURATE DEFINITION OF THE VARIATIONS OF A THERMOMETER

§1.

From the time when the use of barometers and thermometers was discovered, meteorology has been enriched by the great number of observations made by these instruments in so many [different] locations; nevertheless the art of predicting the future state of the atmosphere is still so shrouded in darkness that, judging from the slow progress which has hitherto been made, the whole of eternity would barely be sufficient to advance that skill to the stage where the temperature of the atmosphere could be predicted even a few days in advance. I have often marvelled how it is possible that the courses of stars at such vast distances [from us] have taught men, albeit puzzled [men], in so many centuries to foretell events; while, on the other hand, they have not even begun to discover the art of forecasting changes in the atmosphere surrounding us. Yet this is manifest to all the senses, and the complexity of these changes is not greater, but in fact much less [than those associated with astrological predictions]. Nevertheless, the paradoxical features of this situation disappeared when I considered [the fact that] astronomers in perfecting their skill have adopted a method entirely different from that which meteorologists have followed. For astronomers, after they had seen that there were definite unique periods for any apparent stellar motion, tried first of all to define what they call the 'mean' position of a star, taking no account of inequalities or anomalies. When this had been discovered and compared with the true position, they could then bring to light first of all the greater (more pronounced) inequalities, then also the smaller (more complex) ones, and thus at length arrive at a knowledge of the most intricate movements.

§2.

However, it is evident from the [literature of the] subject itself, that this method has been unrecognized and certainly unused in meteorology, although the method is undeniably no less suitable for meteorology than for astronomy. It is certain that changes in the state of the atmosphere take place according to fixed laws of nature, and are thus restricted to their own periods, just like variations in stellar motions; accordingly it is not impossible to determine both the magnitudes and periods of these changes. We must not imagine that, because of the multitude of causes that have to be taken into account, the subject is so difficult that we cannot in the meantime separate the more noticeable effects from the rest, which are more obscure. So far meteorologists have concentrated merely on observations alone, which, although they can be of great help to this science, do not yield a completely satisfactory conclusion when taken just by themselves. These observations fall into the same category as the huge volume of astronomical observations of, for example, Tycho Brahe;[174] for the motions of stars in future ages cannot be predicted from these [data] alone, unless some Kepler[175] comes along who, in striving after a more profound theory, discovers the laws and magnitudes of the motions. Similarly, meteorological observations, however numerous and accurate, cannot give a positive indication of the future state of the atmosphere.

§3.

Therefore, at this point, we may transfer the example of that astronomical method to variations in the atmosphere, applying it in particular to the ratio of heat and cold: it will thereby be possible to show how I deem that meteorological observations ought to be treated, so that richer fruits can be expected from them.

§4.

In the first place, then, it is well-known that in any one given location on the Earth's surface there is a definite fixed temperature which attains a mean value between summer and winter, and also between day and night, and this depends on the climate, or the elevation of the pole. Just as the rays of the Sun strike the Earth's surface more obliquely, the more they are remote from the equator, so too must this mean temperature be less in places of higher latitude. Moreover, it can easily be shown that if, according to the scale of any thermometer, the mean temperature under the equator is taken as $=m$, and under the poles as $=m-n$, the mean temperature at any latitude ϕ will be approximately $=m-n \sin^2\phi$;[176] or the difference between the mean temperature under the equator and that which obtains at the place whose latitude is ϕ is in a duplicate ratio to the sine of the latitude. Therefore, after ascertaining the mean temperature under two separate latitudes, we can deduce the values of the letters m and n, and accordingly the mean temperature at any latitude whatsoever.

§5.

There can, however, be a different temperature at the same latitude in low-lying and higher places; and in order that outside factors may not influence this value which we call mean and disturb it, places agreeing as far as possible in all particulars, including latitude, must be chosen for the purpose of defining the quantities m and n; of such a kind are coastal places or those whose altitude above sea level is small and equal. Then it will also be appropriate to adapt this mean temperature to these lower places, and to use some correction or formula for higher places, which will show how much this value for the lower place should be diminished to obtain the mean temperature of a higher place in the same latitude. We shall next discuss this correction.

§6.

Having made these remarks I shall, in order to provide an example to illustrate what I have been saying, take the mean temperature in low-lying places near the equator as 24 according to the scale of a Réaumur thermometer.[177] Immediately on comparing the thermometer observations made at the latitude 49° or 50°[178] I discovered that the mean at these latitudes was 9 degrees above freezing point, the same amount in both cases; hence, since $\sin 49° = \frac{75}{100} = \frac{3}{4}$, we should have in the formula set out above

$$9 = 24 - \frac{3^2}{4^2} n,$$

or

$$n = 26\tfrac{2}{3}$$

But the letter n indicates the difference between the temperatures at the equator and the poles; wherefore, subtracting it from the degree 24, the mean temperature at the poles will be $-2\frac{2}{3}$ below freezing point.

§7.

Since, however, I did not take every precaution, or make absolutely accurate observations in [arriving at] this conclusion, it is quite possible that if everything were more carefully examined, with attention also being given to the corrections set out below, the mean temperature at the poles would be found to be exactly 0, or that below which water begins to freeze. Let us suppose, then, for the time being, that both m and $n = 24$, and we shall obtain the mean temperature at any latitude ϕ, expressed in the form:

$$24\,(1 - \sin^2 \phi), \quad \text{or} \quad 24 \cos^2 \phi, \quad \text{or even} \quad 12 + 12 \cos 2\phi$$

and hence we arrive at this table, which exhibits the mean temperature on a Réaumur thermometer at particular altitudes:

[TABLE 1]

Latitude of the place	Degrees on Réaum. therm.	Latitude of the place	Degrees on Réaum. therm.	Latitude of the place	Degrees on Réaum. therm.
0	24	30	18	60	6
5	$23\frac{3}{4}$	35	16	65	$4\frac{1}{4}$
10	$23\frac{1}{4}$	40	14	70	$2\frac{3}{4}$
15	$22\frac{1}{2}$	45	12	75	$1\frac{1}{2}$
20	$21\frac{1}{4}$	50	10	80	$0\frac{3}{4}$
25	$19\frac{3}{4}$	55	8	85	$0\frac{1}{4}$
30	18	60	6	90	0

Moreover, the same formula can easily be adapted to any other scale, whose graduation corresponds to [that of] the Réaumur [scale]. Thus, e.g., if anyone prefers to use the Fahrenheit scale, since 24 degrees on the Réaum. corresponds to around 84 on the Fahrenheit,[179] and freezing point on the latter is 32, it follows that the mean temperature at the elevation of the pole $\phi = 84 - 52 \sin^2 \phi$ or $58 + \cos . 2\phi$ whence this table results:

[TABLE 2]

Latitude of the place	Degrees on the Fahrenheit Scale	Latitude of the place	Degrees on the Fahrenheit Scale	Latitude of the place	Degrees on the Fahrenheit Scale
0	84	30	71	60	45
5	$83\frac{1}{2}$	35	67	65	$41\frac{1}{2}$
10	$82\frac{1}{2}$	40	$62\frac{1}{2}$	70	38
15	$80\frac{1}{2}$	45	58	75	$35\frac{1}{2}$
20	78	50	$53\frac{1}{2}$	80	$33\frac{1}{2}$
25	$74\frac{1}{2}$	55	49	85	$32\frac{1}{2}$
30	71	60	45	90	32

§8.

And it is not inappropriate to compare this mean temperature of any given place with the mean position of planets in astronomy. Moreover, just as astronomers are wont to refer a great number of equations to this mean position, and thus approach nearer and nearer to the truth; so, in the same way, must we investigate the corrections by which the mean temperature can be accurately found.

§9.

Mention has already been made in the preceding paragraphs [§5–7] of the primary correction which depends on the altitude of a place above sea level. Even in the hottest parts of the torrid zone the mountains are perpetually covered in snow or ice at a certain altitude, and this is sufficient indication that when heat rises from the lowest level of the Earth's surface, it gradually decreases; in such a manner, however, that it begins to increase again by a small amount once a certain altitude has been passed. It was French mathematicians[180] who made the valuable observation that a minimum temperature, certainly not greater than freezing point, in the Peruvian territories near the equator predominates at an altitude of around 2400 Parisian toises. In European regions the same [amount of] heat, or rather the same [amount of] cold, is felt at an altitude of around 1200 toises.[181] Therefore, in order that this first temperature correction may be reduced in some way or other to a norm and to a general law, I have proposed in the interim, until the truer law is revealed by the comparison of more observations, that the temperature decreases uniformly with increasing height above sea level, or in proportion to the altitude; and I then established this rule, whose use is very simple.

§10.

If, given the elevation of a place above sea level and its geographic latitude, we have to discover the mean temperature of that place, we must first seek the mean temperature for the given latitude according to the Réaumur scale, either through the formula given above, or from the table set out in the same place [§7]; then a hundredth part of the altitude of the place expressed in Parisian toises, which accounts for this primary correction, must be subtracted from that temperature, and the remainder will be the required mean temperature of the given place.[182] Thus, e.g., for our Göttingen, whose latitude is $51\frac{1}{2}°$ and its altitude above sea level around 70 toises, we must subtract $\frac{70}{100}$ or $\frac{3}{5}$ of one degree from the mean temperature $9\frac{3}{5}$ of the preceding table [Table 1] which corresponds to the latitude $51\frac{1}{2}°$; the remainder will indicate 9 degrees on the Réaumur scale, which is the mean [temperature] of this town.[183]

§11.

The temperature defined in this way occupies an intermediate position between winter and summer, or rather between all [seasons] of the whole year. Having ascertained this either by the above-mentioned rule, or by observations extending over several years (which is preferable for the present case), we can then offer a definite opinion on the temperature of the atmosphere and on the mildness or severity of the climate at a given place, and we can do so with much greater certainty than was possible from those vague and excessively inconsistent maximum and minimum temperatures, which the great majority of people are content to record. Thus, e.g., if the mean temperature for a place is observed to be only 7 according to the Réaumur scale whereas, on the basis of the above rule, it ought to correspond to 9 degrees, that will indicate that the temperature of the air at that place is more severe than usual; and then it is simpler to enquire into the cause of that circumstance. Moreover, there is no doubt that if the temperatures of many places in the world were defined by this method, an increasing amount of light would then be shed in future on physical geography, and indeed even on meteorology and agriculture.

§12.

Another [temperature] fluctuation which we must now discuss is that particularly noticeable one which recurs annually, and divides the whole year into spring, summer, autumn and winter. There are three aspects of this which must be given particular attention: its period, the times of the maximum variations, and the maximum value of the fluctuation itself. Concerning the period, there can be no doubt that it is annual, and recurs on the Earth by virtue of the Sun's [orbital] motion with regard to the cardinal points of the ecliptic; therefore there is no need for us to dwell on this.

§13.

It seems to be more difficult to define the times of year at which the maximum and minimum temperatures occur. Indeed it is well known that both times, at least in these lands of ours, do not coincide exactly with the solstices themselves or with the longest and shortest days, but [occur] a little afterwards. But precisely how great this interval is, we have hitherto been unable to determine satisfactorily—neither by theory, although many attempts have been made, nor even by observations. In general, it seems possible to assume that this interval between the solstices and the times of either extreme of temperature is greater in places which are farther from the equator, and less in places nearer the equator. I leave to others the task [of] investigating this accurately, resting content to set forth the method by which it can be converted to some other use. Thus, for example, I propose that the maximum and minimum temperatures in our lands occur when the Sun is 30° from the solstices, approximately the same interval as thermometric observations suggest; so these particular times correspond to 20 July and 20 January.[184] For other places at a greater or smaller distance from the equator, I shall define this interval in the way noted on this table:[185]

[TABLE 3]

Latitude	The Sun's degrees after a solstice	Latitude	The Sun's degrees after a solstice	Latitude	The Sun's degrees after a solstice
0	0	30	14	60	42
5	½	35	18	65	46½
10	1	40	23	70	50
15	3½	45	28	75	52½
20	6	50	33	80	54
25	9½	55	37	85	55½
30	14	60	42	90	56

Henceforth, I am supposing that the mean temperature occurs when the Sun is removed just that many degrees from the equinoxes and that the variations for any given time of year are in proportion to the sine of the distance of the Sun from the degree of the ecliptic added to this mean temperature;[186] and indeed these hypotheses can hardly differ very much from the truth.

§14.

The maximum value of this annual variation is as yet no less uncertain, since there is the question of whether that theory would be generally valid for all regions of the Earth. But one thing is certain: this variation, or difference between summer and winter temperatures (which comes to the same thing), is greater in regions nearer the pole, but smaller in places further away from the pole. Moreover, by comparing observations taken in this climate of ours, between latitudes 50 and 55 degrees, I discovered that the greatest variation according to the Réaumur thermometer was around 10 degrees, by which [amount] both the summer and winter temperatures are liable to differ from the mean temperature. Hence, having chosen a probable hypothesis without difficulty,[187] I drew up this table which shows the maximum annual variation of temperature at any [given] latitude.

[TABLE 4]

Latitude	Maximum (annual) variation	Latitude	Maximum (annual) variation	Latitude	Maximum (annual) variation
0	0	30	6½	60	11
5	1	35	7½	65	11½
10	2	40	8½	70	12
15	3½	45	9	75	12½
20	4½	50	10	80	12½
25	5½	55	10½	85	13
30	6½	60	11	90	13

§15.

With the help of this and the preceding table, and also with the assistance of that hypothesis which states that the annual variation is proportional to the sine of the distance of the Sun from a point of the ecliptic corresponding to the mean temperature, we can find by calculation the appropriate temperature for individual days in the year. To give an example, I shall construct a table below in this way, for temperatures throughout the whole year; taking the mean temperature as 8, the maximum annual variation as 10 degrees [Réaumur], and the interval between the solstices and times of extreme temperatures as around 30 days.[188]

[TABLE 5]

Months	Days		
	1	11	21
Jan.	−1½	−2	−2
Feb.	−2	−1½	−0½
Mar.	+0½	1½	3
Apr.	4½	6¼	8
May	9¾	11½	13
June	14½	15½	16½
July	17½	18	18
Aug.	18	17½	16½
Sept.	15½	14½	13
Oct.	11½	9¾	8
Nov.	6½	4½	3
Dec.	1½	+0½	−0½

§16.

We shall consider briefly the third temperature inhomogeneity, which similarly is periodical, with a period of one day. It has been observed that the maximum temperature of any given day is attained after noon, and that the minimum occurs at sunrise; it seems, therefore, that the mean daily temperature discovered by the method given above, must be placed at around 9 a.m. and 9 p.m. However, on account of the inequality of days in lands remote from the equator, these diurnal limits of mean temperature must be rather vague. Therefore we must take pains to define these also by means of observations, more accurately than has hitherto been done.[189] Moreover, the magnitude of this diurnal variation —both the maximum and hourly values—must be investigated and established. I shall present my attempt at this in this table which is in front of you [Table 6], as far as the lack of observations and the want of a universal law have allowed me to make an attempt [at all].[190]

[TABLE 6]

Length of day	Hour of the day a.m.							p.m.					
	0	2	4	6	8	10	12	2	4	6	8	10	12
8	−3	−3	−4	−4	−4	−1	+2	+4	+3	+1	−1	−2½	−3
10	−2½	−3	−4	−4	−2	0	+2½	+4	+3½	+1½	0	−2	−3
12	−2	−3	−4	−4	−1	+1	+3	+4	+4	+2	+1	0	−2
14	−1½	−2½	−4	−2½	−0½	+1½	+3	+4	+4	+2½	+1½	+½	−1½
16	−1	−2	−4	−1	0	+2	+3	+4	+4	+3	+2	+1	−1

§17.

Thus, when we have found the mean temperature of any given day from the preceding table [Table 5], we can easily find with the assistance of this table [Table 6] the temperature at any given hour of the day. For example, let the required degree on the Réaumur thermometer be that which should exist in this climate of ours on 10 August at 4 p.m. From the foregoing table, which is appropriate to this climate, it is seen that the mean temperature for the given date (10 August) is 17½; and from this table, since the length of the day is approximately 14 hours, it is found that a correction of 4 degrees must be added at 4.0 p.m.; thus the required temperature will be 21½ degrees [Réaumur].

§18.

I am quite ready to concede that the temperature determined by this method may be vastly different from that which is actually observed, since besides the causes already considered there are others, e.g. winds, clouds, rain etc., which influence heat; but I am far from suggesting that this subject must be considered unworthy of further investigation; on the contrary, I think that it is impossible to define the causes and quantities of the remaining, more involved, variations, unless the effects of the different causes are analysed in the way which I have roughly sketched out here.

II. ASTRONOMICAL OBSERVATIONS TAKEN ON THE MURAL QUADRANT IN THE OBSERVATORY AT GÖTTINGEN

The observations which I have shared with you at different times in this assembly, gentlemen, have hitherto been few in number; the lack of instruments in our astronomical observatory at that time did not permit me to furnish any more, since they would perhaps have been too vague and therefore useless. But now I can set forth a greater amount of observations; not, I hope, of lesser worth by virtue of their quantity. For, around the beginning of this year I gained an opportunity of [using] an outstanding instrument—indeed the best of all thus far invented—and from then on I began to make slightly greater efforts, which demand a more continuous series of observations. And of course I made it my particular object to establish more accurately the positions of the fixed stars, especially of those which are contained in the Zodiac, and to compile a new catalogue of them, superior both in numerical accuracy and in the multitude of stars [comprising it]. After Flamsteed,[191] many have undertaken this task anew, and some have even completed it, so that these efforts of mine (whatever their worth) might easily appear useless. However, when I noticed so many and such great differences in the catalogues which exist, so far was I from deciding to rest content with these catalogues that I was rather strengthened in my resolve to unravel the subject from the beginning and with proper observations.

The instrument which seemed most suitable for this purpose is the mural quadrant, made of brass by the famous craftsman, John Bird of London.[192] In radius it measures six London feet and it has been constructed with great skill both from the point of view of observational accuracy and the convenience of those using it. Three different kinds of divisions are seen on its limb. The first and innermost contains ordinary degrees of which 90 comprise one quarter of the circumference of a circle; each of these is subdivided into five minutes by very fine little lines, and also, by the aid of a device, which the inventor calls a nonius, into 30 separate seconds. Smaller arcs may be estimated by visual judgment, which indeed is not difficult when a microscope is placed over [the scale]. Ninety-six parts of another division, which I shall call the middle one because of its position, are contained in the quadrant of the circle. Each of these parts is separated into 16 smaller ones by slender points, but with the help of a screw, serving the purpose of a micrometer, one can almost perceive the separate seconds. The third division is similar to the middle one except that, in place of the points, lines mark off the intervals. By Nonius's method, the smaller intervals (of which there are again 16 parts, or nearly 13 seconds) and consequently their ten subdivisions, may be obtained by means of visual judgment and a microscope.[193] I use only the outer division, because it is more accurate than the inner one and is easier to use than the middle one, although perhaps this latter could appear preferable because it does not require a visual estimate. However, in observing the culmination of stars, there is for the most part hardly as much time left over from the transit of one star to the transit of the next one as is required to twist the micrometer screw and to count the little parts which it reveals. Thus for that reason I have only used this middle division sparingly, and then [merely] to detect the errors of the outer division, if any doubt should have arisen about it, rather than to record the observed zenith distances.

After this instrument had been secured to a wall in the observatory constructed of special stone, which supported the whole [weight] of it, I began in February of this year to observe the meridian transits of stars and their zenith distances; moreover, I did not neglect the culminations of the Sun, in order that we might further note the true facts concerning this [phenomenon] and the right ascensions of the remaining stars from the true equinoctial point by Flamsteed's method, which is the most reliable of all. In continuation of this task, I measured out the whole circle of the Zodiac and the fixed stars contained in it, which numbered more or less 1000, with the result that there is almost no noteworthy star which has not been observed three or four times at least.[194]

While I was devoting my attention to these subjects, I had at the same time to discover the errors to which the instrument could be subject, both as regards distance from the zenith and the times of transit. And first of all I inquired into the size of the whole arc, to see whether its 90 degrees measured a right angle exactly. Then, with the vertical radius of the quadrant placed with the help of a plumb-line in the true and proper position, the horizontal radius, whose ends were marked with points by the inventor with this aim in view, was investigated with careful accuracy. This was done repeatedly, and on different days, and in every case the latter radius was discovered to be parallel to the horizon as long as the former faced the zenith. From this it can be concluded beyond doubt that the whole arc has been allotted the true quantity which it ought to have, and that the observed zenith distances need no correction on this account.

Then I had to investigate whether the fiducial line of the viewing telescope which is in place of a diopter arm, is parallel to the line passing through the centre of the quadrant and the index; or whether that line points accurately to the zenith while the index is aligned with the zero of the scale on the limb. It was realized that this could best be discovered by turning the instrument round and making a double observation of one and the same star, not very far from the pole; the astronomers who practice this usually call this procedure 'the rectification of the instrument around the zenith'. And so on the 21 July, after the zenith distances of certain stars in Draco had been observed during the preceding days, the quadrant was turned from the meridional region, which it had faced until then, towards the north, in order to observe the same stars. When this was done, a fault was discovered in the cylinder around which the tube revolves. For the screws to which the body of the quadrant should be firmly fixed, were found to be loosened—I don't know why I imagine that this is the reason why, before this defect had been noticed, the distances from the zenith of stars near it were found to be five seconds greater than afterwards, when the fault had been corrected. On 6 August the quadrant was restored to its former position, i.e. facing the meridian, and once more the stars just mentioned were observed. From this series of observations, which I shall subsequently designate by (*a*),[195] it is clear that the zenith distances taken with this instrument appeared greater by only two seconds, which are [inherent] in the instrument itself, and thus this error can henceforth be eliminated from the observation of [such angular] distances.

As regards inequalities of divisions on the limb, these certainly either do not exist at all, or are no greater than those tiny errors which are considered negligible for all practical purposes because they escape perception by the human senses, no matter what aids are employed to help them [i.e. the senses]. I know that several people have suggested a long pair of compasses to explore the intervals of degrees (but I find that nowhere have they themselves actually followed up this suggestion). For my own part, however, I would not put much stress on this experiment, which seems too unreliable and suitable only for discovering larger errors that a skilled craftsman can easily avoid in the very operation of dividing; on this account I have disregarded this experiment entirely. Those three arcs mentioned above, of which each is completely separate from and independent of the others, show wonderful agreement in providing the most definite proof of a division properly completed on this instrument of ours: they show this in response to my repeated experiments, and moreover on different points of the limb. The sole discrepancy which appears to anyone making this comparison, is that the inner division is always a distance of 4·5″ from the zenith more than the outer; but since this discrepancy is the same everywhere, it does not arise from the inequality of the parts, but rather from the fact that the indices, or beginnings of each division, do not lie in one and the same straight line with the centre of the instrument.

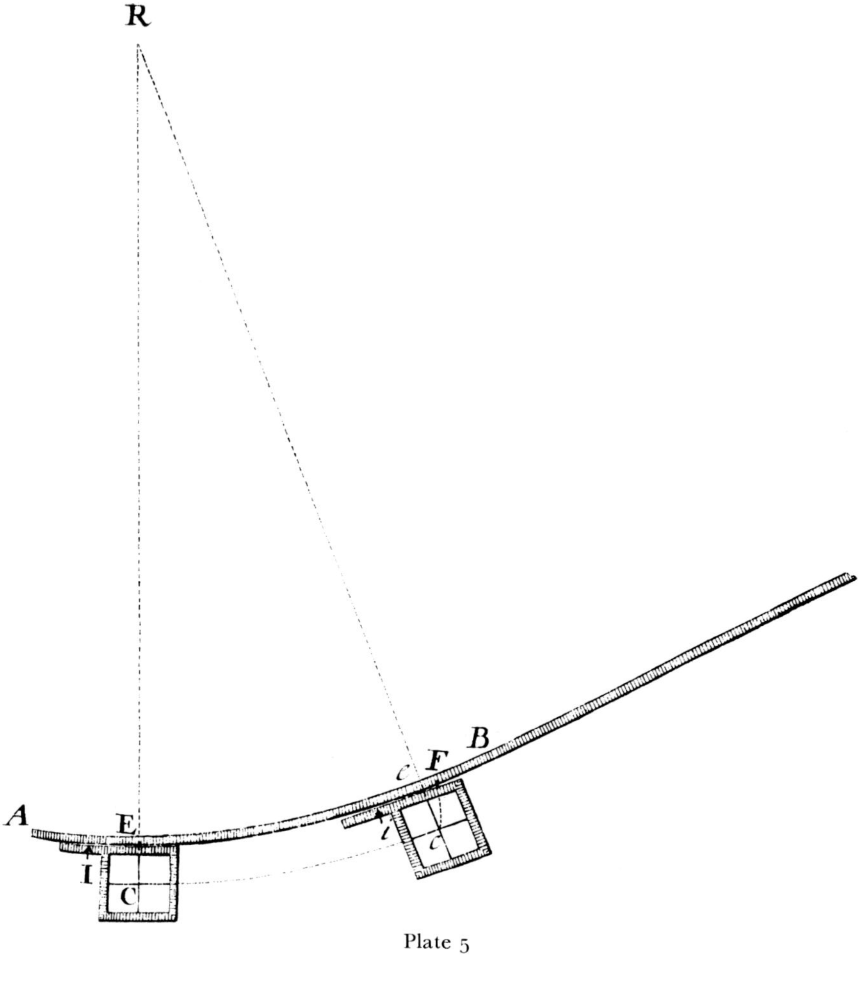

Plate 5

Moreover, these are for the most part errors which astronomers usually take into account in instruments showing zenith distances or altitudes. For it is commonly supposed that these instruments have been properly examined and adapted for practical use (rectified, as others say), if we are assured about: the size of the whole arc; the equality of the parts on the limb; the parallelism of the fiducial line with the line joining the zero of the scale and the centre; and finally, about the position of the centre which in this instrument of ours (at least after the above-mentioned error has been corrected), the rotating telescope tracing out circles exactly parallel to those which are on the limb, shows to be precisely located. However, I have discovered another type of error, besides these ones, which as far as I know has been noticed by no one up to now in instruments of this type;[196] and unless we guard against this error, or at least investigate it, it can prove no less harmful to accuracy in observations than any of those errors thus far mentioned—especially since it is far more difficult to deduce the true value of this error for any given point on the limb.

This error can occur in all those instruments in which the telescope moves over the limb. In this case, it is impossible that the fiducial line or axis of the telescope should fall on the actual surface of the limb without its inevitably tracing out by its own movement a surface [which is] indeed parallel to it, but at a distance as great as that between the limb and the intersection of the cross-wires at the focus of the telescope. Hence it follows that if the limb is not perfectly even, this fiducial line or rather the intersection of the cross-wires will pass through either a greater or smaller space than that which in the meantime the index has traced out over the limb itself, according as the limb is convex or concave in that place. Let us suppose in the figure Tab. I (Plate 5) that AB is the defective part of the limb, so that on the side where the tube falls, it is convex; C is the point on the focal plane of the telescope where the cross-wires intersect; and I is the index. When, therefore, the intersection of the wires C on c in the telescope is accomplished by tracing out the path Cc parallel to the limb AB, the index over the limb traverses the line Ii. Moreover, when perpendiculars CE, ce, are dropped from C and c on to the limb, the result is IE; i.e., and thus the path Ii traced out by the index is equal to the distance Ee; but from the nature of the curved parallels this is shown to be less than Cc. Hence it may easily be gathered that, for this reason, the zenith distances shown by the instrument are less than the true distances, and indeed by the quantity eF, since cF is parallel to CE. Similarly, the opposite could occur if the limb were concave anywhere, i.e. it [the error] would produce zenith distances greater than in actual fact.

But in a matter so manifestly obvious, I hardly think that there is need for me to remind you that this error is quite different from that which the curvature of the limb might perhaps produce by itself, by a contraction of the whole arc, and which would exist even if the fiducial point C were to fall on the very surface of the limb. About this error which is of no great importance unless the curvature is enormous, I shall, as I say, make no [further] mention here, but I shall deal instead with that which is caused by the distance of the point C from the limb combined with the curvature of the limb. Here, even, if the limb diverges the slightest fraction from perfect uniformity, that fraction is nevertheless not so slight that it can be disregarded, as one example will suffice to demonstrate. Let us suppose that ER and eR are the normals drawn from any two points E and e on the limb, intersecting at R; hence, since cF is parallel to CR, the following will be approximately true: ce : eF = Re : eE, that is the error eF = eE(ce/Re). Hence, if eR is 10 000 times greater than ec, and the arc Ee = 30 degrees, the error eF = 108 000″/10 000 = 11″ approximately. Moreover, since on our instrument the distance of the point c is around $1\frac{1}{4}$ inch from the limb, the point R, which can be regarded as the centre of curvature of Ee, is more than 1000 feet distant from the limb; and the limb, if the curvature is taken as circular, diverges only by a thousandth part of a foot from a perfect plane at the mean position between e and E.[197] Anyone who is experienced in these matters will readily see how difficult it is to guard against such a tiny curvature of the limb, from which, nevertheless, an error of 11 seconds in the zenith distances can arise.

That the curvature of the limb in our mural quadrant does not amount to this quantity is, however, clear from the fact that the error in the zenith distances nowhere equals 11 seconds. I am forced to infer from the deviations of the limb from the meridian plane, of which I shall speak at greater length later, that it is quite another [order of magnitude]. The movable quadrant was suitable enough for recording the altitudes which are called corresponding,[198] yet because of its defect I could not explore these deviations of the limb, nor its curvature, to my complete satisfaction. However, all the rest of the investigations thus far set in progress prove sufficiently conclusively that the limb at around 24 degrees from the zenith is slightly convex, from the thirtieth to the sixtieth degree it is sensibly plane, and then it is slightly concave.[199] I have tried to determine the actual value of the error arising from this, or at any rate the approximate value; and in order that the following observations may be freed from that error, it will be worthwhile to set them out in this table (Table 7) along with that which was discovered through stars culminating at the zenith.

You will see from this table, gentlemen, what great accuracy the instrument possesses, when tiny errors not even equal to three seconds (for greater errors nowhere occur) were not able to escape detection. But I

[TABLE 7]

Zenith distance	Error	Zenith distance	Error	Zenith distance	Error	Zenith distance	Error
°	″	°	″	°	″	°	″
−5	−2·2	20	+1·0	45	+2·6	70	+0·0
−4	−2·2	21	+1·2	46	+2·6	71	−0·1
−3	−2·1	22	+1·4	47	+2·6	72	−0·2
−2	−2·1	23	+1·6	48	+2·6	73	−0·3
−1	−2·0	24	+1·8	49	+2·5	74	−0·5
0	−2·0	25	+2·0	50	+2·5	75	−0·6
1	−1·9	26	+2·1	51	+2·5	76	−0·6
2	−1·8	27	+2·2	52	+2·4	77	−0·7
3	−1·7	28	+2·3	53	+2·3	78	−0·7
4	−1·6	29	+2·4	54	+2·2	79	−0·7
5	−1·5	30	+2·5	55	+2·1	80	−0·7
6	−1·4	31	+2·5	56	+2·0	81	−0·7
7	−1·3	32	+2·6	57	+1·8	82	−0·6
8	−1·3	33	+2·6	58	+1·7	83	−0·6
9	−1·2	34	+2·6	59	+1·5	84	−0·5
10	−1·1	35	+2·6	60	+1·4	85	−0·5
11	−1·0	36	+2·6	61	+1·2	86	−0·4
12	−0·8	37	+2·6	62	+1·1	87	−0·3
13	−0·7	38	+2·6	63	+0·9	88	−0·3
14	−0·6	39	+2·6	64	+0·8	89	−0·2
15	−0·5	40	+2·6	65	+0·7	90	−0·1
16	−0·3	41	+2·6	66	+0·5		
17	−0·0	42	+2·6	67	+0·4		
18	+0·3	43	+2·6	68	+0·2		
19	+0·6	44	+2·6	69	+0·1		

must remind you that this table serves to correct only those zenith distances which were observed after 21 July. Errors which affected the instrument before that date may be more safely ascertained by a comparison of the zenith distances recorded before and after its conversion, than by any other means.

Having discussed the errors of the zenith distances, we must now deal with those which can affect the meridian transits observed by this instrument. The origin of these is two-fold. First of all, since the times of transit are noted from a clock, whose regular motion may be continually disturbed by many causes, the observed transits may differ from the true ones for the very reason that they are based upon a regular motion.[200] For this reason too, they [the observed transits] stand in need of some correction, because it is hardly possible to ensure that the individual points of the limb are distributed along the whole meridian, or that the cross-wire coincides everywhere with its plane.

In my observations, I have strived to narrow the bounds and reduce [the magnitude of] the errors of the first kind, which cannot be reduced to fixed laws because of the multitude and unreliability of factors producing them and therefore cannot be considered as known [quantities]. I used simultaneously two clocks of very delicate construction and compared, whenever necessary, the motion of the one which was placed at about 90 degrees [literally, a quadrant] from the observer's line of sight, with the motion of the other [placed] 30 feet away, so that at an intermediate position, the beat may be distinctly perceptable to both ears and the amount of difference between them easily distinguishable to the tenth part of a second. In this way the errors of clocks which arise from the friction of the [cog-] wheels, the irregularity of the teeth, and the viscosity of the oil with which they are lubricated, are for the most part detected and corrected. But it is impossible to ascertain in the same manner those errors which the daily temperature variation produces by changing the length of the pendulums, since the general effect of this cause in clocks of an identical construction, as mine are, is the same. However, these errors may for the most part be avoided if in recording the right ascension, or its difference, from the observed transits of stars, the time of the daily rotation [of the Earth about its axis] is derived from observations at an interval of 24 hours, but not in fact from any more widely separated ones. In my diary, I have very frequently noted down observations of this kind, from which this interval can be found. However, on the rare occasions when the transit occurs precisely at some stroke of the pendulum or at an exact second of time, I have expressed the fractions of seconds to be added in decimal parts, both because it is quicker to write these down, and also because through long use and practice it comes more naturally to me to calculate tiny units of time according to these [as opposed to sexagesimal] fractions.

It remains for me to say something about the other error; that is, about the deviations of the limb from the meridian plane. In all, three causes combine to produce these discrepancies, besides the curvature of the limb from which our mural quadrant is not free, as I mentioned above. The first of these causes is the inclination of the quadrant plane, or its deviation from the vertical plane, which means that, although the radius has been fixed exactly horizontally in the meridian, yet in every other position the telescope stays outside it—and this is greatest around the vertex [or zenith]. If the quadrant suffered from this defect only, the times of the observed transits might easily be referred to the meridian itself, the deviation on one point of the limb being known; for in general it is proportional to the cosine of the zenith distance divided by the cosine of the declination from the equator, i.e. when the declination $=\delta$, the zenith distance $=\alpha$, then the deviation will be $=A\cos\alpha/\cos\delta$, A denoting a constant quantity, for whose determination a single observation will clearly be sufficient. But if there is also another cause to be taken into account, namely the variation of the horizontal radius from the meridian line, then an error is produced [which is] proportional to the sine of the zenith distance divided by the cosine of the declination from the equator; and the entire deviation, arising from both causes taken together, can be expressed by $(A\cos\alpha+B\sin\alpha)/\cos\delta$. In addition, however, a third cause can be incurred, which happens if the fiducial line [i.e. the axis of the telescope] is not parallel to the plane of the quadrant, but traces out a conical surface. The error in the times of the culminations which is produced in this manner is inversely proportional to the cosine of the declination. The complete deviation arising from these three causes combined may be expressed by this formula $(A\cos\alpha+B\sin\alpha+C)/\cos\delta$.[201]

Thus, having discovered the deviation on three different points of the limb, the value of the quantities A, B and C may be determined; and then, using this formula, the deviation for any given point of the edge may be found—provided that the limb is perfectly regular. And in turn, once the deviation in more than three points is known, the curvature of the limb can be deduced; this will, in fact, be zero if the deviations observed agree with the formula, but the contrary will be true if it is impossible to make the formula agree with all the deviations observed.

Following this method, I have therefore discovered some curvature on the limb of our mural quadrant, although I have not yet ascertained with absolute certainty its true amount at any given point on the limb. But some movement in the wall to which the instrument is fixed greatly hampered me in this investigation; it was slight, indeed, but nevertheless in the course of time it caused the silver thread by which the quadrant is held in a vertical position, and which in February was hanging very freely from the limb, to hold it [i.e. the quadrant] quite tightly by April. And although I readjusted the position of the quadrant in this same month, so that the suspension might not touch the limb to any greater extent, it nevertheless came nearer again afterwards, and did not come to rest before July. I gathered from these observations that the wall had gradually leaned over, either because its foundation had not yet properly subsided, since it was only recently constructed, or because the rock from which it is built did not until then lose, by the summer heat, the moisture which it had absorbed in the stone-quarry. At length this movement ceased entirely, and from August right up to the present time I have not been able to detect even the slightest dislocation of the wall. In the meantime, however, as I have already indicated, that movement prevented me from being able to determine as yet the deviation of the quadrant from the meridian plane and consequently the curvature of the limb, with anything approaching accuracy. Wherefore, in what follows I am obliged to suppress these observations, which cannot be dealt with unless the deviation of the quadrant is known; I shall communicate them on another occasion, after I have satisfied myself on this matter.

What I have been saying are words of advice to those who desire to employ the following observations (a)[202] for astronomical uses. For I think that it is the duty of every observer to give a public account of the condition of his instrument, of the care and precaution which has been taken to treat it properly, and of the methods of inquiring into its errors, so that others too have the opportunity of judging how much reliance is to be placed in the conclusions drawn or to be drawn from our observations. Since astronomers of a former generation did this less, treating their investigations too carelessly and theoretically, the obvious result was that their observations quickly became obsolete, when compared with such vigilance and such great labour.

III. A SIMPLE AND ACCURATE METHOD OF COMPUTING SOLAR ECLIPSES VISIBLE IN A GIVEN PLACE

In calculating the phases of solar eclipses the ancients unquestionably used for the calculation of the parallax, a method both complex and ungeometric, and accordingly astronomers of a more recent age have tried in many ways to make this process easier and more accurate. Among them the very learned Kepler was the first to see that solar eclipses can be regarded as obscurations of the Earth, and we can scarcely exaggerate the extent to which the universal doctrine of solar eclipses has been illuminated and perfected by that discovery of his. Soon that old approach was abandoned, when a consideration of terrestrial eclipses revealed another way which did not require that highly complicated calculation of the [solar] parallax. By conceiving that at the time of any solar eclipse the Earth is seen from a long way off, it was possible by means of optical precepts and the [method of] projection which is called orthographic, to represent quite accurately and easily the various phenomena that are observed from [literally, that affect] the Earth's surface during an eclipse. It was even permissible to omit most of the calculations, since the graphical operation took their place. Equally, those things which ought to be seen on the Sun with an eye fixed on the surface of the Earth—the beginning, middle, end, and magnitude of the eclipse etc.—could be learnt with the same ease from this projection.[203] Thus the majority of the difficulties seemed to have been removed by this fortunate and praiseworthy invention; except that geometric accuracy, which is of primary importance, had at the same time suffered a considerable setback from the invention.

For, as often happens, the invention was well received because of its novelty and simplicity, so that what should have served only to elucidate the theory of eclipses, was eagerly seized upon as the norm and rule, and no one could be found who was concerned about its geometric demonstration. How else could it have come about that the many defects, by which it is certain that this application of the projection was hampered, are still unknown and that authors, who have undertaken to describe and expound it in their writings, disagree in many particulars, with none of them having in the meantime arrived at the actual truth?

It seems hard that a procedure which for a whole generation was adopted by almost everyone, stands convicted of falsity and error, with no reason being brought forward. Therefore, let us indicate one error of the projection; an important one, to which it is not easy to apply a remedy. No one acquainted with this theory can be unaware that the position of the eye, which is chosen for the purpose of viewing the Earth at the time of any eclipse, is placed on a line drawn through the centre of the Earth and the Sun, and that the circles of the Earth—namely, the meridian, horizon and parallel of that place, to which we intend to refer in particular the phenomena of the eclipse—are then projected on a plane normal to this line; also, that the ecliptic projected on to the same plane, and the path of the lunar shadow, are conceived of in the same way—both of them as straight lines. Although certain aspects of this [method] are clearly not consistent, I do not propose to investigate them separately now. However, I cannot by any means give my consent to the fact that the periphery of the shadow projected by the Moon is represented on the same plane as if [it were] circular. That this is false and that the shape of the shadow is quite different from a circle can be demonstrated thus: the curves which are generated by a section of a cone made by a plane, are known (among other things) as conic sections. However, if the surface cutting the cone is not even, but bent, e.g. spherical, the section must be a curve of another type, which geometricians call a line of double curvature. Moreover, it is certain from optics that this curve, whatever its nature, will appear as a perfect circle to an eye fixed on the vertex of the cone; but, on the other hand, it nowhere appears circular if the eye is placed beyond this vertex—except, perhaps, for a few cases, which there is no room to discuss in the present treatise. Now let these principles be applied to the projection of terrestrial eclipses. The shadow projected from the opaque body of the Moon is a cone of the same kind; and when a spherical or spheroidal surface of the Earth intercepts it, the resulting section and curve of both is entirely of double curvature, which nevertheless would appear circular if the eye were imagined as having been placed on the vertex of the shadow cone; that is, on some point of the line joining the centre of the Moon and the Sun. Since this is not the case on the common projection, inasmuch as it necessarily requires the position of the eye [to be] on the line drawn through the centre of the Earth and the Sun, and moreover infinitely remote, it is clear that the shadow is incorrectly represented as being circular, and that that projection which is so much praised clearly lacks any geometrical basis.[204]

If anyone should think that this defect in the projection can be corrected or eliminated either by changing the position of the eye, or by defining more accurately the shape of the projected shadow, I shall not contradict him; but I am afraid that, when one error is removed, other new ones may be introduced in greater quantities, and that the precision and simplicity by which this type of projection has hitherto commended its use may be lost. Moreover, anyone who would say that these defects are not great enough to warrant our ceasing to use the projection would be acting against his own interests, since he would be showing himself to be ignorant of the state of more recent astronomy, which tries to avoid much smaller errors than these, and has also to some extent succeeded in avoiding them.

For my part, I am quite convinced that astronomy can easily—nay, rather, with some profit even—do without that apparatus of projection. It is possible to make the calculation of eclipses considerably easier, or at least clearer, by the introduction of an orthographic projection; although we have not yet reached the stage where complexity and trouble have been entirely removed from the subject, as is clear from the precepts of astronomers which are adapted to orthographic projection, e.g. those which are found in the German edition of the Hiraean tables.[205] Anyone following these precepts in calculating the beginning, middle, end and magnitude of any solar eclipse, requires to establish no fewer than 24 analogies according to trigonometry, and even these are not sufficient to determine exactly the instants mentioned unless 12 others are added. Anyone who says that this is less work, must certainly never have approached the matter seriously.

Nor is the method of projection together with the principles of its use which the distinguished de la Caille published in the Memoirs of the Royal Academy of Sciences Paris in 1744 any easier, but rather more complicated, although he thinks that he has corrected the calculation of projection and freed [it] from even its tiniest errors.[206] Apart from the fact that he supposes the shape of the Earth to be exactly spherical, his method suffers from the same defect which I have shown above to be latent in every projection. The same points must clearly be noted with regard to the method of the distinguished Klingenstiern included in the Acts of the Royal Scientific Society of Upsala in 1742, p. 106,[207] since it is not built on any more solid foundation, nor is it free from the disadvantages and obscurities of so many calculations, as is abundantly clear even from a single glance at the accompanying example. I therefore believe that we should leave aside the orthographic projection [and] the consideration of eclipses of the Sun as obscurations of the Earth, which, as has hitherto been shown, does not rest on any geometric foundation and is not very convenient for formulating the actual computation, and adhere more to the concept which the ancients devised for themselves regarding these eclipses. What is simpler and more in accordance with the nature of the event than if we think of the beginning of the eclipse as taking place when the limb of the Moon is first seen to touch the edge of the approaching Sun (that is, when the distance of the centres is seen to be equal to the sum of the apparent semi-diameters of the Sun and Moon); the greatest darkness as occurring when the centres of the luminaries are nearest; and the end, when the limb of the Moon receding from the Sun again touches the Sun's limb and the apparent distance of the centres is again equal to the sum of the semi-diameters? In this way the whole problem is reduced to an investigation of the apparent motions of the Moon and the Sun, such as they are at the time of any given eclipse; when these are known, it is quite easy to discover those instants when the distance of the centres or of the whole orbs is least, or equal to the sum of the apparent semi-diameters of Sun and Moon.

Here in a few words, then, is a straightforward path and the most natural of all, which I think we must follow in calculating solar eclipses, and it does not differ from that followed by the ancients except that certain devices, to be described in detail later, have been added, with the help of which the difficulties and deficiencies which were troublesome, are removed or diminished. I shall set out systematically the precepts which this method suggests, and illustrate them in the appended example.

First, I propose that before everything else the day and also the approximate hour of the approaching eclipse be recorded in some way. Then I conceive of three equally-spaced instants of time of which one, whose value may be predetermined, is close to the beginning of the eclipse, the second close to the middle, and the third close to the end; and at these instants I compute from the theoretical tables the true longitude and latitude of both the Sun and the Moon, and the horizontal equatorial parallaxes of the semi-diameters. The calculation of all these things cannot be more lengthy than the calculation which is carried out in the customary manner to discover the time of the true new-moon ecliptic.

Secondly, I seek the altitude of the Moon above the horizon at any one of these given instants of time; not by calculation, but from an artificial globe, or by some other cruder method. For this altitude is used only to find the apparent semi-diameter of the Moon from the given horizontal [one], which ultimately suffices to express it in whole degrees at any rate. Moreover, the reduction of its diameter to that altitude is contained at different places in the accompanying tables, under the title of the increase of the horizontal semi-diameter of the Moon, and for that reason there is no need for any calculation here. However, if anyone should wish to provide this [information] by calculation, without any prior knowledge of the altitude of the Moon, he may do so satisfactorily by employing this general formula.[208]

The increase of the semi-diameter of the ☽

$$= \frac{+D^2(\cos\phi\cos\delta\cos\mu+\sin\phi\sin\delta)}{56\,254}$$

in which D [is] the horizontal semi-diameter of the Moon, ϕ the altitude of the pole, δ the declination of the Moon, and μ its distance measured from the meridian along the equator.[209] Moreover, it should be noted that the sign is positive if the declination is northerly [and] negative if it is southerly.

Thirdly, from the elevation of the pole at the place to which the calculation of the eclipse is directed, I subtract the correction corresponding to that in the second column of the following table, and the remainder I call the reduced elevation of the pole. In the same way, I subtract from the equatorial parallax the correction from the third column of the same table, and what remains is the horizontal parallax of the Moon, and indeed of the said place on the horizon. Moreover, these corrections arise from the spheroidal shape of the Earth, and it is by these alone that we can obtain entirely satisfactorily the conclusion which this departure of the Earth from a perfect globe demands in the calculation of the parallaxes which has subsequently to be undertaken. From this starting point, since the table is very easy to use, the computation of the eclipses is not at all complicated.

[TABLE 8]

Elevation of the pole	Correction of the elevation of the pole		Correction of the equatorial parallax if the parallax were 54′	60′
[ϕ°]	′	″	″	″
0	0	0	0·0	0·0
5	2	59	0·1	0·1
10	5	53	0·5	0·5
15	8	36	1·1	1·2
20	11	3	1·9	2·1
25	13	10	2·9	3·2
30	14	53	4·0	4·5
35	16	10	5·3	5·9
40	16	56	6·7	7·4
45	17	11	8·1	9·0
50	16	56	9·5	10·6
55	16	10	10·9	12·1
60	14	53	12·1	13·5
65	13	10	13·3	14·8
70	11	3	14·3	15·9
75	8	36	15·1	16·8
80	5	53	15·8	17·5
85	2	59	16·1	17·9
90	0	0	16·2	18·0

Fourthly, at the elevation thus reduced and at those particular instants I investigate the ninetieth [degree] of the ecliptic,[210] and its altitude or angle of rising, using either the special table constructed for that elevation of the pole, or more general tables which are everywhere available, for example, in Strauch's extant tables of oblique ascensions and of angles of rising,[211] or in Gruneberg's most popular edition of the canon of sines.[212] But, in the absence of tables of this kind, the following formulae will show the ninetieth [degree] and its altitude. Given the reduced elevation of the pole $= \phi$, the obliquity of the ecliptic $= \alpha$, the right ascension of the middle of the heavens (or of the culminating point of the equator, which may easily be deduced knowing the time and the right ascension of the Sun) $= \mu$, the angle ω must first of all be sought. When this is found from

$$\tan \omega = \frac{\tan \phi}{\sin \mu}$$

$$\tan \text{nonag.} = -\frac{\tan \mu \cos (\omega - \alpha)}{\cos \omega}$$

$$\cos \text{alt. nonag.} = \frac{\sin \phi \sin (\omega - \alpha)}{\sin \omega}$$

Moreover, it is sufficient if in this calculation both the given and the required arcs are expressed in degrees and minutes, for any error arising from the omission of seconds in parallaxes rarely exceeds one second.

Fifthly, once the nonagesimal point has been ascertained, together with its altitude, and hence the true distance of the Moon from it, I set out to find the parallaxes of the longitude and latitude of the Moon from the formulae proposed by myself in Vol. II of the Transactions of the Royal Society [of Sciences, Göttingen];[213] according, however, to these following [formulae], which are slightly better suited to the present use and purpose.

Given the horizontal parallax of the Moon $= \pi$
the altitude of the 90th [degree of the ecliptic] $= A$
the true distance of the Moon from the nonagesimal $= b$
the true latitude of the same $= a$
the sine of one second $= s$

it follows that

the parallax in the longitude $= \pi \sin A \sin b + s\pi^2 \sin^2 A \sin b \cos b$
the parallax in latitude[214] $= \pi \cos A \mp \pi \sin A \sin a \cos b$
$\pm s\pi^2 \sin A \cos A \cos b$

In order that we may at the same time take account of the solar parallax, it will be suitable in these formulae to substitute in place of π the difference between the horizontal parallaxes of the Sun and Moon; the rest may be completed like the other [formulae].

Sixthly, when the parallaxes have been discovered, I add them to the true positions of the Moon, so that it consists of its apparent motion, both absolute and relative to the Sun. I set out in a table the difference of the apparent longitude of the Moon and Sun, and also the apparent latitude of the Moon for three single instants, and by employing interpolation I extend it to five or ten minutes of the intervening time, in [the course of] which I write down at the same time the sum of the semi-diameters of the Sun and Moon. These things having been disposed of,

Seventhly, by comparing the observed latitude and the differences of longitude with the sum of the semi-diameters, I estimate what latitude approximately corresponds to the beginning and end of the eclipse which, although it does not yet consist entirely of these instants, cannot be difficult since often the latitude scarcely changes by a minute in a whole hourly interval; moreover, it is sufficient for the purpose to have foreknowledge of it to within this limit. Therefore, I call this imaginary latitude x, and the corresponding sum of the semi-diameters from the table, c, and I seek the quantity $\sqrt{c^2-x^2}$; and if in the table between the differences of longitude the required quantity has a corresponding latitude equal to the imaginary one, that will indeed be the true difference of longitude at the beginning and end of the eclipse. If, however, there is a discrepancy between these latitudes, let it be called dx; let $\sqrt{c^2-x^2}$ be to x as dx is to the fourth term which, added to $\sqrt{c^2-x^2}$ if the imaginary latitude is too great, or subtracted from it, if it is too small, will yield the correct difference of longitude in accordance with the beginning or the end [of the eclipse]. Wherefore, if the corresponding times are obtained from the table, the actual instants of both the beginning and the end may be found.

Eighthly, to discover the time of the maximum obscuration, I take the hypothetical latitude or the one approximately true at that time as being $=x$, its increase or decrease around the same corresponding to the interval of five or ten minutes $=dx$, the increase or decrease of the difference of the longitude for the same interval $=dy$; and I make $dy : dx = x$: the fourth term y, which will be the difference of longitude at the time of the maximum eclipse. The table will readily show the actual time, only be careful to note that y must be sought within the differences of longitude before apparent conjunction if the latitude is increasing, and vice versa.

Ninthly, since, therefore, the true time of the greatest obscuration is given, the truer latitude corresponding to this time will also be given from the table; and if it is now substituted for the hypothetical one, the shortest distance between the centres will be $\sqrt{x^2+y^2}$. When this is subtracted from the sum of the semi-diameters the result converted into sixths of the Sun's semi-diameter, will appear as the maximum quantity of the eclipse.

[TABLE 9]

EXAMPLE

Calculation of a Solar Eclipse

on 26 October 1753 for the horizon and meridian at Göttingen, according to my lunar tables inserted in Vol. II of the transactions.[215]

The instants of time substituted for the true time on 26 October a.m.	9h 20′ 0″	10h 30′ 0″	11h 40′ 0″
Given that the meridian difference between Paris and Göttingen = 30′ 16″, from the tables there will follow:			
the true longitude of the Sun	7s 3° 6′ 19″	7s 3° 9′ 14″	7s 3° 12′ 9″
the true longitude of the Moon	7 1 58 36	7 2 39 43	7 3 20 49
the true northern latitude of the Moon	27 49	31 37	35 24
the semi-diameter of the Sun	16 10	16 10	16 10
the equatorial parallax of the Moon	59 8	59 7½	59 7
the horizontal semi-diameter of the ☽	16 7½	16 7	16 7
The altitude of the Moon around the increase of the half Moon	3½	6	8
the apparent semi-diameter of the ☽ in that altitude	16 11	16 13	16 15
the sum of the semi-diameters of the ⊙ and ☽	32 21	32 23	32 25
The elevation of the pole at Göttingen	51° 32′	51° 32′	51° 32′
Reduction from the table	16	16	16
Reduced elevation of the pole = ϕ	51 16	51 16	51 16
Correction of the equatorial parallax from the table	10	10	10
Horizontal parallax of the ☽	58′ 58″	58′ 57½″	58′ 57″
Distance of the ⊙ eastwards from the meridian	40° 0′	22° 30′	5° 0′
Right ascension of the ⊙	210 53	210 56	210 58
therefore the right ascension of the middle of the sky = μ	170 53	188 26	205 58
hence, given the obliquity of the Ecliptic 23° 29′ the nonagesimal of the ⊙ =	4s 26° 47′	5s 9° 52′	5s 23° 56′
the altitude of the nonagesimal = A	47 29	41 15	34 27

[TABLE 9]—*continued*

The true distance of the ☽ from the nonagesimal $=b$	65° 2′	52° 48′	39° 25′
the horizontal parallax of the ⊙	0′ 11″	0′ 11″	0′ 11″
hence the difference between the horizontal parallaxes of ⊙ and ☽	58 47	58 46½	58 46
and the parallax of the longitude of the ☽	39 29	31 5	21 16
—the parallax of the latitude	39 48	44 16	48 32
Therefore the observed longitude of the moon	7 2 38 5	7 3 10 48	7 3 42 5
The difference of the observed longitude of the ⊙ and ☽	west 28 14	east 13 4	east 29·56
The observed southern latitude of the ☽	11 59	12 39	13·8

Hence the following table is completed by the rules of interpolation; in it those intermediate instants of time have been omitted, since there is no use for them here.

[TABLE 10]

Time			Observed longitude difference		Observed southerly latitude		The sum of the semi-diameters	
h	′	″	′	″	′	″	′	″
9	10	0	−32	36	11	52	32	21
9	20	0	−28	14	11	59	32	21
9	30	0	−23	54	12	5½	32	21
10	20	0	−2	36	12	34	32	23
10	30	0	+1	34	12	39	32	23
10	40	0	+5	42	12	44	32	23
10	31	0	+25	58	13	4½	32	25
11	40	0	+29	56	13	8	32	25
11	50	0	+33	52	13	11	32	25

Now put the visible latitude around the beginning $=x=12'\ 0''$, and the sum of the semi-diameters $c=32'\ 21''$, then $\sqrt{c^2-x^2}=30'\ 2''\frac{1}{2}$. This difference of longitude indicates that the beginning falls almost half way between $9^h\ 10'$ and $9^h\ 20'$; at which time the truer observed latitude=

11′ 56″, differing 4″ from the hypothetical one. Therefore, as 30′ 2″½ [is] to 12′ 0″ so is 4″ to the fourth term 1″½, and when this is added to 30′ 2″½ it gives the true difference of longitude for the beginning = 30′ 4″. Since, moreover, the decrease of the difference of longitude between 9^h 10′ and the beginning = 32′ 36″ − 30′ 4″ = 2′ 32″, and the same decrease between 9^h 10′ and 9^h 20′ = 32′ 36″ − 28′ 14″ = 4′ 22″, the result is that 4′ 22″ [is] to 10′ 0″ as 2′ 32″ [is] to the fourth term, which will be 5′ 48″.

Similarly, given the latitude of the Moon at the time of the end of the eclipse = 13′ 0″ and the sum of the semi-diameters = 32′ 25″, then $\sqrt{c^2 - x^2}$ = 29′ 42″. Moreover, according to the table, the truer latitude 13′ 8″ corresponded to this difference of longitude, differing 8″ from the hypothetical one. If, then, it is the case that [as] 29′ 42″ [is] to 13′ 10″, so 8″ [is] to 3″½ and if this 3″½ be taken from 29′ 42″, the remainder will be the truer difference of the longitude at the time of the end of the eclipse = 29′ 38″½, and thus, using a proportional part, as was done shortly beforehand for the beginning, the time of the end will be discovered to be 11^h 39′ 13″.

Let the latitude at the time of maximum obscuration = 12′ 30″ and since, according to the table, for ten minutes around this time the increment of the longitude difference dy = 4′ 10″ and the increment of the latitude = 5″, the fourth proportional to 4′ 10″, 5″ and 12′ 30″ is required, which will be = 15″. This required factor is found among the negative differences of longitude, on account of the increasing latitude, so the time of greatest obscuration corresponding to the proportional part, [is] 10^h 25′ 38″.

Finally, since at this very time the observed latitude of the moon according to the table = 12′ 37″ = x, and the difference of longitude as before = 0′ 15″ = y, the nearest distance of the centres will be = $\sqrt{y^2 + x^2}$ = 12′ 37″½, and when this is subtracted from the sum of the semi-diameters 32′ 23″, the maximum amount of the eclipse remains = 19′ 45″½ which, in digits of which six are equivalent to the semi-diameter of the Sun 16′ 10″, are worth 7^d 20′.

See the observation of this eclipse made by me at Göttingen in Vol. III of the Transactions of the Royal [Scientific] Society [of Göttingen], p. 453.[216]

VI. TREATISE ON THE RELATIONSHIP OF COLOURS

§1.

When I embarked on a discussion of the number, strength and mixture of colours, it appeared to me that there was a basic fact to notice amidst such a variety of colours:

There are three colours, simple or primary, and three only—red, yellow and blue. From a mixture of these all other colours are produced, but they themselves cannot in any way be produced from other colours, no matter how the other colours may be mixed together. We see them in the rainbow, and even more clearly in a sunbeam captured in a glass prism, although there they appear together with and surrounded by other secondary colours. Some people count these secondary colours among the primary ones, so that they name seven primary colours (why not more, when in truth they are infinite?) red, orange, yellow, green, blue, indigo, violet;[217] but since they do not explain sufficiently clearly why they want these colours to be included in the category of primary colours, truly nothing can be deduced from their opinion [which is] contrary to my own way of thinking. If the definition of primary colours is taken to be as follows: all other colours derive from a mixture of them, but they cannot by any means be produced from any colour combinations, we shall easily convince those whose opinion differs on this subject; for experiment has proved that no other colours, besides those three named, can be cited, to which these characteristics pertain.[218]

§2.

I do not count white and black among colours properly so called, although they commonly go under that name. For white may be more correctly compared to light, black to shade or darkness; and natural philosophers do not deny this, considering white as a perfect mixture[219] of all colours, and even as a ray of light, but black as the absence of colour, i.e. darkness. But I shall expound my opinions on these two colours at greater length later on.[220]

§3.

As regards the secondary colours, some of them are mixed from only two primary colours, others from three, and these mixtures differ among themselves according as one or other colour predominates. Since the proportions of primary colours in a mixture can be varied infinitely, the colours produced are themselves infinitely diverse. I can enumerate the main ones only, by way of example.

§4.

From the mixture of red and yellow the following colours are produced: saffron, gold, orange, then flame red and a ruddy colour—the former if there is more yellow in the mixture, the latter if there is more red. When they are mixed in equal proportions, the colour produced can most fittingly be called red–yellow, since it is difficult to allot single names to single colours.

§5.

Yellow and blue mixed together produce greens of every hue. The colour consisting of equal parts of yellow and blue is that which is most strictly speaking green: it shades off from green into yellow when it contains more yellow, but if more blue is mixed with the yellow, a sea-green colour is produced.

§6.

A mixture of red and blue produces purple, and strictly speaking that name can only be applied to the colour which contains equal parts of red and blue. Pink, however, has more red in it, violet more blue.

§7.

These, then, are the colours which are produced from a mixture of two primary colours. They can all be seen in a rainbow, and in a sunbeam intercepted in a prism. There, ruddy shades of every hue, or colour tones from red to yellow gleam between red and yellow; and greens of every hue between yellow and blue. Moreover, red on the one side and blue on the other, passing through various grades, end up as purple or violet.

§8.

But those colours which are produced from a mixture of the three primary colours no longer appear in a rainbow or in a refracted ray of sunlight. For a ray of light itself, if physicists are to be believed, is a mixture, as it were, of the three primary colours. But what its colour is you will not find easy to define. White has been extracted from a white substance, red from a red one, and yellow from a yellow one, and perhaps no colour corresponds to its own self, but is more correctly merely called the cause of colours. But this is not the place to introduce intangible problems which, although they could be unravelled, have nothing to do with the subject in question. Whatever Newton asserts to the contrary, one thing is certain: when red, yellow and blue are mixed, neither light nor white are produced, but dark browns, chestnuts, reds, greys and ashen colours of every shade, depending upon the relative proportions of the primary colours included.

§9.

When white is mixed with either primary or composite colours, it produces no new colours, but only renders them paler, and for this reason it can hardly be classified as a colour. However, if anyone prefers to count these pale colours as new ones, I shall not contradict him, if he adds white to the colours; for it is senseless to quibble over mere words. Black behaves in the same way—when it is mixed with other colours, it does not create new colours, only darker ones. Indeed, individual colours can be called black if they are very dark. But generally, black is considered to be only that colour which, when mixed with white or toned down in some other way, becomes more ashen-coloured. In this case, black must be counted among the composite colours, since it becomes more ashen-coloured when red, yellow and blue are mixed together in due proportion.*

§10.

When I am speaking about mixing colours, it is quite clear that I am not referring to those artificial mixtures or chemical infusions such as that, for instance, by whose aid the black of our writing ink is generally produced. It is well known concerning many substances that their natural colours can be removed by the addition of a certain acid, or, rather, can be changed into other colours. In a few, even the action of air has the same effect. In the present treatise I do not include mixtures produced by violent reactions, and transmutations of this kind. Therefore if anyone wants to test for himself by experiments the truth of what I have said about producing composite colours, he must use pigments which contain no acid that might act on the colours of the pigments after they were mixed. It will be safest to employ pigments which, [having been] reduced to dust, can be mixed dry.[221]

§11.

Although the proportion of the quantity of primary colours in a mixture can be varied in innumerable ways, and accordingly must yield innumerable different colours, all these colours are nevertheless not equally distinct. If, for example, only a twentieth or a thirtieth part of blue were mixed with yellow, the resulting colour might counterfeit some appearance of green, but it would differ so slightly from yellow itself that merely by

[* Its formula, according to the notation described below, is $r^3g^2b^7$].

looking, one can hardly distinguish it from yellow—indeed, it could not be distinguished from it at all. Colours of this type, which can only with difficulty be distinguished from neighbouring shades, we shall call, with all due respect to them, indistinct.

§12.

In order that we may have some fixed rule, if colours of this kind are to be produced which are distinct from others, it will be convenient to follow the same rule which architects and musicians have drawn up for themselves. It will no doubt be observed that the proportion of colours in a mixture is such that it cannot be expressed by large numbers. For the smaller these numbers are, the more distinct are the colours which emerge, mixed from primary shades in varying ratios. Both in architecture and in music numbers greater than twelfths are not readily admitted, since their ratio can scarcely be perceived by the unaided senses; so too, in this harmony of colours (a name which it will not be inappropriate to apply to the whole art of painting) we may stop [sub-dividing, or grading] at the number 12. In this way we shall have between any two primary colours 11 intermediate composite colours, which can still be distinguished by the naked eye; the rest, falling between these, we shall rightly count among the indistinct colours.

§13.

It will be convenient to observe this rule not only in the case of the colours deriving from two primary colours, but also in the case of those which are composed of all three primary colours. Once it has been laid down—a thing attempted by no one as yet and which at first sight may seem impossible—we can proceed to enumerate all the different colours individually, and reduce them to a fixed scale; so that, given any colour whatsoever, it will immediately be possible to tell by what ratio it was created from the primary colours. For the sake of convenience, we shall signify red by the letter r, yellow by g[222] and blue by b: composite colours we shall define by juxtaposing the letters of the primary colours from which they are mixed, so that, for example, the formula for the colour mixed from red and yellow will be rg or gr; [that] for the colour consisting of the three primary colours rgb, etc. The proportion in which any colour is composed of primary colours, we shall express by numbers placed above the letters, like exponents, so that if a colour consists of three parts of yellow and two of blue, its formula will be g^3b^2. I need scarcely remind you that numbers written thus are not really of the same kind as those which correspond to the exponents of the algebraists. Wherefore, to distinguish them by name, let us call them *distributives*.

§14.

It is obvious that if the distributives of any colour have a common factor they [still] signify the same colour if divided by this factor, since only the ratio of these distributives, not their absolute magnitude, is taken into account in colour mixing. Thus, e.g. r^8g^4 signifies exactly the same colour as r^2g^1. Conversely, if individual distributives are multiplied by the same number, you must not think that the colour changes. Thus in general, any distributive must be regarded as the numerator of a fraction whose denominator is the sum of all the distributives which occur in the same expression of colour: i.e. r^8g^4 is the same as $r^{8/12}g^{4/12}$, etc., but for the sake of brevity the denominator is omitted.

§15.

In enumerating, therefore, the individual colours which differ by not less than a twelfth part of the whole both from primary colours and from other composite ones, the entire problem is reduced to [that of] investigating the number of ways in which the number 12 can be arrived at from combining two or three smaller integers, and how often the sequence of the small letters rg, gb, rgb, can be permuted. This problem is not difficult to solve. When the combination is made, the following colours emerge:

(1) Primary colours:

r^{12}	g^{12}	b^{12}

(2) Colours composed of two primary colours:

$r^{11}g^1$	$g^{11}b^1$	$r^{11}b^1$
$r^{10}g^2$	$g^{10}b^2$	$r^{10}b^2$
r^9g^3	g^9b^3	r^9b^3
r^8g^4	g^8b^4	r^8b^4
r^7g^5	g^7b^5	r^7b^5
r^6g^6	g^6b^6	r^6b^6
r^5g^7	g^5b^7	r^5b^7
r^4g^8	g^4b^8	r^4b^8
r^3g^9	g^3b^9	r^3b^9
r^2g^{10}	g^2b^{10}	r^2b^{10}
r^1g^{11}	g^1b^{11}	r^1b^{11}

(3) Colours composed of three primary colours:

$r^1g^1b^{10}$	$r^2g^1b^9$	$r^3g^1b^8$	$r^4g^1b^7$
$r^1g^2b^9$	$r^2g^2b^8$	$r^3g^2b^7$	$r^4g^2b^6$
$r^1g^3b^8$	$r^2g^3b^7$	$r^3g^3b^6$	$r^4g^3b^5$
$r^1g^4b^7$	$r^2g^4b^6$	$r^3g^4b^5$	$r^4g^4b^4$
$r^1g^5b^6$	$r^2g^5b^5$	$r^3g^5b^4$	$r^4g^5b^3$
$r^1g^6b^5$	$r^2g^6b^4$	$r^3g^6b^3$	$r^4g^6b^2$

$r^1g^7b^4$	$r^2g^7b^3$	$r^3g^7b^2$	$r^4g^7b^1$
$r^1g^8b^3$	$r^2g^8b^2$	$r^3g^8b^1$	
$r^1g^9b^2$	$r^2g^9b^1$		
$r^1g^{10}b^1$			
$r^5g^1b^6$	$r^6g^1b^5$	$r^7g^1b^4$	$r^8g^1b^3$
$r^5g^2b^5$	$r^6g^2b^4$	$r^7g^2b^3$	$r^8g^2b^2$
$r^5g^3b^4$	$r^6g^3b^3$	$r^7g^3b^2$	$r^8g^3b^1$
$r^5g^4b^3$	$r^6g^4b^2$	$r^7g^4b^1$	
$r^5g^5b^2$	$r^6g^5b^1$		
$r^5g^6b^1$			
	$r^9g^1b^2$		
	$r^9g^2b^1$	$r^{10}g^1b^1$	

§16.

In this list, then, all those colours are included which differ from their neighbours by a twelfth part; certainly barely as much as can be perceived by the sense of sight. No colour can be produced which does not have its exact counterpart among these, so that the graduation never amounts to more than a twenty-fourth part, which indeed is imperceptible. We can therefore assert with the utmost confidence that every single different colour is included in that table. Therefore their overall total is as follows:

Primary colours	3
Colours composed from two primary colours	33
Colours composed from the three primary colours	55
Hence, all the different colours that exist	91

§17.

Since this number is triangular and its side is 13, all the colours can be fitted into a distinct triangle in 91 little areas in such a way that the three primary colours are its angles, colours composed of two primary colours are its sides, and those composed of three occupy the central area, being nearer to the primary colours the more they derive their hue from them. Table II (Plate 6) illustrates this arrangement.

§18.

If those individual areas are actually filled in with the colours which are denoted by the formulae written in them—quite an easy task for those who know how to use pigments—a kind of scale of colours will be obtained, with the help of which any colour at all can be discovered and called by its own name, just as varieties of gold and silver are mixed on a Lydian

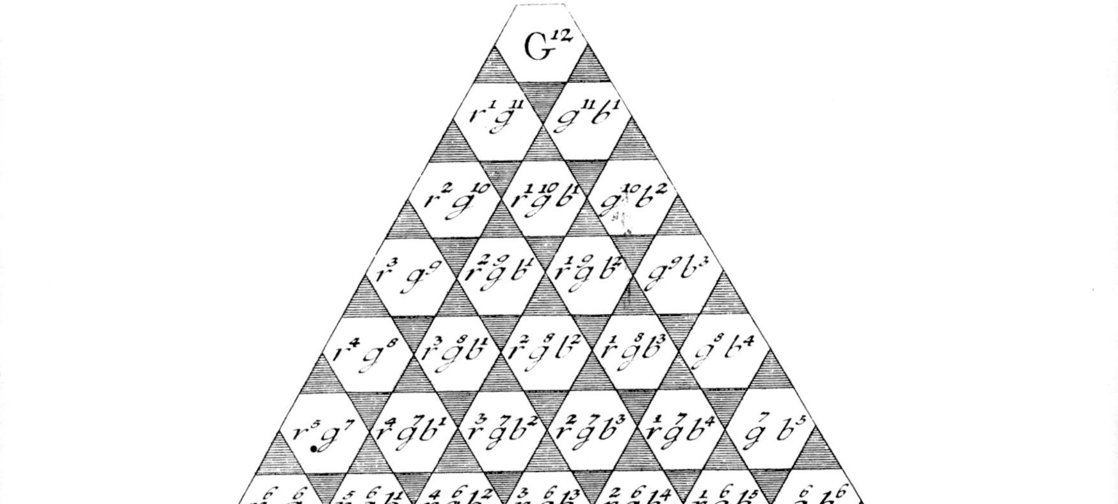

Plate 6

stone. At the same time the method according to which a colour can be mixed from the three primary colours may be described.

§19.

A scale of colours of this kind will have many applications, especially in painting where, since pigments corresponding to individual colours do not exist, it is mostly extremely difficult to produce any given colour from a mixture. Up to now, artists seem only to have succeeded by trial and error. However, provided that individual pigments are pure and still unmixed with any others, when once their relation to the primary colours has been defined with the help of the scale mentioned above, it will be easy to manufacture a suitable pigment from a mixture of others, since any given colour does not have its own pigment. Conversely, it will be possible to describe what sort of colour will be produced if certain pigments are mixed together in a given ratio.

§20.

The matter may be clarified further by examples. Let us suppose that the following formula has been found to agree with the scale of colours:

r^{12}—a pigment made from cinnabar [viz., vermilion],
$r^{8}g^{4}$—a pigment made from red lead,
$r^{4}g^{6}b^{2}$—a pigment made from yellow ochre,
$r^{1}b^{11}$—a pigment made from Berlin blue;

and in a similar manner the formulae for the rest of the pigments are established. Someone may ask what sort of colour will result if three parts of cinnabar-red, two of Berlin blue, and four of yellow ochre are mixed. To answer this question, let the distributive symbols of any given pigment be multiplied by the number of the parts contained, thus:

red lead	yellow ochre	Berlin blue
$r^{8}g^{4}$	$r^{4}g^{6}b^{2}$	$r^{1}b^{11}$
3	4	2
$r^{24}g^{12}$	$r^{16}g^{24}b^{8}$	$r^{2}b^{22}$

then let the distributives of each letter be added thus:

$$r^{42}g^{36}b^{30}$$

and divided by the number of ingredient parts $3+4+2=9$, and the colour sought will be: $r^{4\,2/3}g^{4}b^{3\,1/3}$ which, however, since the distributives are fractional, is contained among the indistinct [colours]; nevertheless, it will be very close to the distinct [colour] $r^{5}g^{4}b^{3}$, i.e. it consists of five parts of red, four of yellow, and three of blue.

§21.

Another example. Let us take a colour whose formula is $r^7g^2b^3$, and let us inquire how many parts of red lead, yellow ochre, and Berlin blue must be mixed to produce the given colour.

Let the parts of red lead $= x$
of yellow ochre $= y$
of Berlin blue $= z$

then, following the preceding example, the resulting colour will be[223]

$$r\frac{8x+4y+z}{x+y+z} \quad g\frac{4x+6y}{x+y+z} \quad b\frac{2y+11z}{x+y+z}$$

Since this must be the same as $r^7g^2b^3$, *it ought to follow that:*

$$\frac{8x+4y+z}{x+y+z}=7 \quad \text{or} \quad x-3y-6z=0$$

$$\frac{4x+6y}{x+y+z}=2 \quad \text{or} \quad 2x+4y-2z=0$$

$$\frac{2y+11z}{x+y+z}=3 \quad \text{or} \quad 3x+y-8z=0$$

Since the sum of the first and second equations produces the third, the solution is indeterminate; but the system of equations is satisfied by $x=3z$, $y=-z$. *Since no physical meaning can be attached to a negative number of pigment parts, one is obliged to conclude that it is impossible to produce the colour whose formula is* $r^7g^2b^3$ *from a mixture of red lead, yellow ochre, and Berlin blue.*

§22.

The problem will be insoluble if it is proposed to mix any colour from more than three pigments; and since this is virtually impossible, no doubt the component parts from the solution of the equations will turn out to be negative. But these and other similar problems many prefer not to pursue, since they follow readily from the associative law.

§23.

So far we have been considering perfect colours, i.e. colours endowed with that degree of light whereby they gleam to their full extent. It remains for us to deal also with those which diverge from this mean—and as it were, proper—state, into paleness or darkness or, if you prefer, into light or shade. Indeed, any colour is rendered pale if it is mixed with white; but white itself is palest of all. Let any standard colour be represented by the expression $r^m g^n b^t$. Therefore, if the letter w denotes white, that colour, made paler by the mixture of white, will be $w^p(r^m g^n b^t)^q$ or $w^p(r^{mq} g^{nq} b^{tq})$, where p signifies how many parts of white, and q how many parts of the colour itself are included in the mixture.

§24.

It must also be observed here as was noted above concerning the perfect colours, that colours mixed with white do not exhibit such a high degree of distinctiveness as the colours themselves. Thus, e.g. although these perfect colours r^{12} and $r^{11}b^1$ differ by one degree of distinction, these same colours, when mixed with an equal number of parts of white, do not differ by the same degree, but only by half [of it], and this is much more difficult to distinguish. Hence, if we suppose that between white and any primary colour there are 12 degrees of distinction, it follows that when one part of white is mixed with 11 parts of red, yellow or blue, the number of intermediate colours, or of colours mixed from these primary colours toned down with white, will be less than the number of colours mixed from perfect primary colours. As the number of these last is triangular, of side 13, it is easily demonstrated that the number of those which contain one part of white, is triangular, of side 12. Similarly, the number of those which contain two parts of white is triangular, of side 11, and so on. This is easily demonstrated if in the general formula set out above of colours mixed with white, $w^p r^{mq} g^{nq} b^{tq}$ are taken as distributives, and p, mq, nq, tq as integers whose sum total is 12. Since it is too much trouble to express the formulae of these colours individually—this can easily be discovered by anyone if necessary—I shall only set out their sum total. Therefore following what has been said, the distinct colours will be:

1 part white contained	$w^1 r^m g^n b^{11-m-n}$	78
2 parts white contained	$w^2 r^m g^n b^{10-m-n}$	66
3 ,, ,, ,,	$w^3 r^m g^n b^{9-m-n}$	55
4 ,, ,, ,,	$w^4 r^m g^n b^{8-m-n}$	45
5 ,, ,, ,,	$w^5 r^m g^n b^{7-m-n}$	36
6 ,, ,, ,,	$w^6 r^m g^n b^{6-m-n}$	28
7 ,, ,, ,,	$w^7 r^m g^n b^{5-m-n}$	21
8 ,, ,, ,,	$w^8 r^m g^n b^{4-m-n}$	15
9 ,, ,, ,,	$w^9 r^m g^n b^{3-m-n}$	10
10 ,, ,, ,,	$w^{10} r^m g^n b^{2-m-n}$	6
11 ,, ,, ,,	$w^{11} r^m g^n b^{1-m-n}$	3
12 or perfect white	w^{12} — — —	1
therefore, the sum total of colours mixed with white		364

where it is relevant to note that this is a pyramidal number, whose side is 12.[224]

§25.

The dark colours, or those verging on black or darkness, remain [to be discussed]. In order to examine more accurately the nature of these [colours], we must first consider that any perfect colour is endowed with a certain fixed degree of light, which renders it most conspicuous. If this light is completely removed from either a primary or a compound colour, it leaves perfect black. If we suppose that some portion only of this light is absent, the resulting colour will be dark, although of the same name. But the more difficult it is to distinguish a colour from others lacking the same portion of light, the greater that portion is; or rather, the colours themselves are darker. If all light were absent, all the colours would be black, and so there would be no difference between them.

§26.

It must also be noted that any dark colour is made brighter by mixing it with white, and indeed so much white can be added that it becomes perfect instead of being dark, or grows paler. Since no colour is perfect unless it can be understood to have been produced in this way from a darker colour, provided that this is of the same name, it hereby follows that in any perfect colour there is a fixed quantity of white which, if it could be separated, would leave the colour either totally or partially black or dark.

§27.

From this I think it is now clear that the reason for light and white producing colour is the same. For if light is taken from colour, or white is separated from it, [the colour] comes out dark or black. But light differs from white in that the former when mixed with any colour that is intrinsically dark, or with black itself, does not make it brighter or bring it nearer to perfection; while the latter, when mixed with black or a darker colour, can produce any perfect colour. Moreover, perfect colours by their nature must necessarily occupy a mean position, as it were, between white and black, and they are produced if equal parts of white and black are mixed together. Wherefore, just as we have counted 12 intermediate degrees from any perfect colour to white itself, so we must assert that there is the same distance from black to each perfect colour.[225]

§28.

Moreover, we can likewise conclude that dark colours, which seem to be mixtures of perfect colours and of black, must rather be considered as mixtures of perfect colours and of negative white. For black is a sort of multifarious or indeterminate colour, since it can be created from any perfect colour. Accordingly, if it is mixed with white or another colour, it produces now one colour, now another. But this ambiguity disappears if we consider black as a privation of white.[226]

§29.

Therefore, just as above we denoted any colour mixed with white by the expression $w^p(r^m g^n b^t)^q$, so shall we most conveniently indicate a colour mixed with negative white, i.e. a colour bordering on black or a dark shade, by the same formula, changing only [the sign of] the distributive *p* of w itself to negative; thus $w^{-p}(r^m g^n b^r)^q$. Hence, as many dark colours exist as there are pale colours, i.e. 364. Finally, it is possible to state the total number of every single distinct colour, whether perfect, or pale, or dark:

Perfect colours	91
Pale colours	364
Dark colours	364
The sum of all will be	819

§30.

Anyone who has paid a little more attention to the great variety of colours, cannot but wonder at the vast scope of the art of painting, which can convert all these colours to a useful purpose. If all languages are formed from a combination of the 20-odd letters of the alphabet; if from the varied scale of musical tones, whose number however rarely exceeds 50, the art of music can harmonize so many different compositions; how much richer and more fruitful must we deem the art of painting, since it has more than 800 colours at its disposal from which to produce its works in differing structure and composition? But it is fitting that the Wisdom by which this world was founded has provided for the most outstanding of the senses with the greatest variety of objects.

V. TREATISE ON THE NEW CATALOGUE OF ZODIACAL STARS

I am glad to bring to you today, gentlemen, the same new catalogue of zodiacal stars which I announced at this meeting two years previously,[227] happily completed at last. A catalogue of around 1000 fixed stars, which would fill a thick volume if all the numbers which have been written to establish it, together with the computations made, were to be included in one book, is condensed into a few pages. Perhaps there will be those [among you] who, in order that they may more clearly judge for themselves the success or otherwise of this task which I have undertaken, wish for the whole series both of observations and computations to be placed before their eyes at the same time; but the procedure of our Society does not permit such comprehensiveness nor, to tell the truth, have I hitherto enjoyed the leisure that would be needed to collate the individual figures in the order in which they could be offered to the public; but I trust that I shall satisfy most people if I set out here the means by which I deduced from observations the positions of the individual stars contained in the catalogue.

The instruments by which the observations of the heavens have been taken were sufficiently discussed when I announced my intention of producing a new catalogue. To recap briefly, the principal instrument is a six-foot [radius] quadrant, made entirely of brass, and fixed to a solid wall in such a way that it is in a very close alignment to the plane of the meridian.[228] With its aid, both the distances of the stars from the pole, and [the times of] their meridian transit were observed. These times were recorded by a pendulum clock of exquisite construction, which during observation was frequently compared with another no less accurate, in order to detect any irregularities which might appear. Changes in the atmosphere which appreciably affect not only refractions but even the timekeeping properties of clocks, were carefully and quickly investigated with a barometer and thermometer. Finally, a 15-inch radius portable brass quadrant, which I myself had divided with the greatest possible accuracy, served to record equal altitudes and to determine the wall quadrant's inclination to the meridian plane,[229] although, as I shall presently demonstrate, a change in position of the zodiacal stars may also be discovered without this, and with an even greater degree of accuracy.

The manner and order in which these observations were carried out is sufficiently clear from what was said on the above-mentioned occasion. Thus there is no need to discuss this matter any further. I merely add this, that a series of observations was begun in February 1756 and continued throughout the whole of two years and more, although it was quite frequently interrupted, both on account of the more turbulent conditions which had meanwhile arisen in this land and town,[230] during which I did not consider it safe enough to go up to the observatory at any hour of the night whatever, and also because I considered it superfluous to watch stars carefully which had already been observed so many times before.

After I had obtained a sufficient number of observations, it was particularly incumbent upon me to employ them to determine accurately the right ascensions and declinations of stars. The following method seemed the most reliable, and the one best suited both to the observations and to the instruments. In the first place, from observations of certain of the brighter stars among those which are contained in the tropical zone and which were observable even during the daytime, I deduced such differences in right ascension as appeared on the plane of the instrument, and having worked out the precession of the equinoxes, aberration and nutation, I reduced them to a common epoch, viz. the beginning of the year 1756. Indeed, these differences would be true and absolute, if the plane of the instrument corresponded accurately with that of the meridian, or if it differed by the same amount at any given point of the zodiac. I then investigated the true right ascension of each one of these stars from a comparison of its transit [time] with that of the Sun, which was moving on roughly the same parallel [of declination] both in its [daily] ascent northwards and in its descent southwards; by applying the corresponding zenith distance of the Sun, the latter's distance from either equinoctial point was found.[231]

In this way, therefore, knowing the right ascension of the Sun, I was also able to obtain [a value for] the right ascension of a star which was accurate and independent of the lack of alignment of the plane of the quadrant [to that of the meridian], since the discrepancy must be the same for the Sun whose motion is parallel to that of the star. But this method of deriving the right ascensions of stars is similar to that which, after Flamsteed[232] and Manfredi,[233] astronomers today usually follow, except that they determine the differences between the right ascensions of the Sun and a star by [comparing] corresponding altitudes; whereas I deduce this information from the transit of both [these heavenly bodies] through the plane of the quadrant, which comes to the same thing provided that the declination of the Sun and the star, or the zenith distance, is the same.

But this method is reliable only if the Sun (and therefore the star also, whose right ascension is required) is close to the tropics inasmuch as its superiority depends entirely on the swiftness of the Sun's motion in declination, which is extremely slow in the vicinity of the tropics. The right ascensions of stars which lie nearer either tropic were for that reason investigated in a special manner. The method is as follows: since the solar motion was known, from the previous method, in all points on the ecliptic except those close to the solstices, it follows that one can discover the errors in the solar tables, and correct the latter so that they provide a sufficiently accurate indication of the Sun's [orbital] motion throughout the whole year, and therefore around the [times of the] solstices also. When this has been done, the true right ascension of the Sun (moving on the parallel of a star in the vicinity of a tropic) calculated from these tables, and the difference between the right ascension of both [Sun and star] simultaneously observed, the right ascension of the star is immediately obtained.

And with these aids I derived the true right ascensions of stars which were particularly bright, working them out accurately and reducing them to the same epoch. I did this not once or twice, but repeatedly, whenever the opportunity presented itself; for I considered that no efforts ought to be spared in order that these stellar positions of those stars, which ought to provide a standard and foundation, as it were, for all the others, should be established as accurately as possible.

Now, since these stars were so situated that each of them crosses the plane of the quadrant at a different distance from the zenith, it was not difficult to examine the individual points of the limb [of the quadrant] and to distinguish whether the deviation [of the plane of this instrument] with respect to the meridian is the same in individual cases, or whether it is different. If the true right ascensions of two stars crossing the meridian at different points of the limb (discovered by the method already described) were found to differ by the same amount as the observed time interval between their respective transits over the plane of the quadrant, we should be entirely justified in concluding that there was no error in alignment at either point or at any rate no difference in the inclination [of the plane of the quadrant with respect to the meridian at either point]. If, however, on the other hand, the difference of the true right ascensions proved inconsistent with [that given by] the observations, it would obviously suffice for the purpose to know how much, in what sense, and in what region [of the sky] one point varied from another, even if the amount of the deviation itself were uncertain; however, in order that I might be in no doubt about the true times of observations in determining the positions of the planets and especially of the Moon, I derived this quantity by [the method of] equal altitudes[234]—for certain points on the limb, at any rate—and from this I obtained the [angle of] deviation of the remaining points of the whole limb, to the extent to which it is contained within the tropics,[235] both through [observations of] the stars, and also by the use of interpolation.

These [quantities] having thereby been established, the positions of the remaining fainter stars still remained to be investigated—a task which required no greater skill, but one involving greater patience and more work. Anyone who knows the length of the calculation needed for investigating precession, aberration and nutation, will readily concede that it is an extremely tedious task to go over the same calculation three or four thousand times when the position of one fixed star is being observed on different occasions; and possibly I would not have had sufficient patience to carry out this task, if certain compendia had not existed by which the greatest part of the task may be removed.[236] It will be worth while to discuss these compendia separately at some other time, since they may be no less useful to those who will use the catalogue than they were to me in constructing it.

But this was only half my task; for as well as the right ascensions, the declinations of the stars had also to be established. An inquiry had again to be made for this purpose into the elevation of the pole which, after repeated observations and determinations, I discovered to be 51°31′ 54″.[237] Then [the values of] refractions and their variations [which were] dependent upon the different state of the atmosphere, had also to be carefully scrutinized.[238] Finally, the errors in the division of the quadrant and in the positioning of the diopter arms had to be investigated. Indeed if I should wish to explain by what means and with what great care all these things have been determined by me, I should far exceed my prescribed limits. What was said by me on these points in the treatise read on 6 November 1756 [Treatise II] may suffice for part at least. But besides these corrections, which are common to all stars, I also computed, for individual stars, the corrections due to the precession of the equinoxes, the nutation of the Earth's axis, and the aberration of light, which affect the declinations,[239] and duly applying these, I ultimately recorded the declinations themselves in the catalogue.

Now I shall add a few words about the catalogue itself and its construction. It contains under separate headings the right ascensions and mean declinations of zodiacal stars referred to the beginning of 1756. At the same time I have added at the end the number for both the right ascensions and declinations, indicating how often each of them has been observed, so that a judgment can be made concerning the degree of certainty with which the position of a zodiacal star may be estimated. For there is less accuracy in the position of one whose right ascension and declination have been observed only once or twice, than in that of those which are observed more often. And indeed I think that a position deduced from only one observation may vary by about 10″, while one deduced from ten observations varies only 2″;[240] the right ascensions of the fixed stars for which no such number is appended, have been deduced from a comparison with the Sun itself, and may be considered as a foundation for the rest. Another number for the magnitude, or brightness index, of a star precedes its name.[241] The region of the declination, whether northern or southern, is indicated by the signs + and −, the former denoting the north, the latter the south.

I have not added any longitudes and latitudes, because I have not so far had as much leisure as their calculation requires, and also because they are rarely used in everyday astronomical practice:[242] moreover, the changes which have been discovered in our time in the obliquity of the ecliptic, in the [celestial] latitude of the stars, and in their very positions, no longer allow us to consider the ecliptic co-ordinates as fixed.[243] But these changes and the proper motions of fixed stars may be more conveniently discussed on another occasion.

Magn.	characteres.	Ascens. recta	Num. Obseru.	Declinatio.	Num. Obseru.
2	Pegasi γ	0. 10. 29, 6	4	13. 49. 37, 9✠	3
7	.	0. 36. 34, 8	1	7. 27. 54, 7✠	1
7	.	1. 0. 54, 1	1	6. 53. 7, 0✠	1
8	.	1. 2. 33, 5	2	0. 29. 39, 5✠	2
7	. .	1. 19. 15, 2	3	0. 19. 53, 1✠	3
6	♓ d	2. 1. 2, 7	4	6. 50. 4, 9✠	3
7	.	3. 0. 38, 6	2	3. 34. 7, 4—	2
6.7	.	3. 13. 43, 0	2	0. 35. 12, 8✠	2
7	.	3. 31. 58, 8	4	1. 24. 6, 9✠	3
8	.	3. 42. 37, 2	2	1. 27. 45, 7✠	2
6.7	.	4. 24. 0, 4	2	5. 18. 22, 8✠	2
6.7	.	4. 57. 30, 7	1	5. 36. 20, 1✠	1
9	.	5. 15. 16, 3	1	1. 57. 15, 2—	1
6.7	.	5. 45. 37, 2	2	1. 50. 49, 4—	2
7.8	.	6. 24. 18, 4	2	1. 50. 47, 0—	2
6	.	7. 4. 30, 6	2	5. 41. 53, 4—	2
9	.	7. 38. 59, 7	1	5. 11. 44, 0—	1
8	.	8. 9. 30, 4	3	1. 4. 51, 9—	3
6	.	8. 42. 2, 5	1	5. 24. 23, 9✠	1
6.7	.	8. 54. 12, 2	1	4. 1. 21, 4✠	2
7.8	.	8. 54. 52, 1	4	5. 57. 50, 5✠	4
4	♓ δ	9. 0. 49, 1	4	6. 15. 13, 8✠	5
7.8	.	9. 41. 32, 4	3	2. 3. 32, 2✠	4
6	.	10. 8. 26, 0	1	2. 28. 20, 7—	2
8	.	10. 11. 47, 8	2	2. 45. 33, 1✠	2
8.9	..	10. 40. 31, 7	1	5. 31. 40, 3✠	1
7	.	10. 50. 46, 4	1	12. 37. 32, 8✠	1
7	.	11. 17. 5, 3	2	12. 22. 21, 0✠	2
7.8	.	11. 25. 26, 7	2	5. 31. 24, 6✠	2
7	.	11. 48. 9, 3	2	5. 9. 43, 9✠	2
4	♓ ε	12. 34. 38, 4	2	6. 34. 21, 1✠	2
8	.	12. 47. 15, 7	1	5. 27. 3, 0✠	1
7.8	.	13. 3. 56, 4	2	4. 20. 37, 7✠	2
7.8	.	13. 18. 27, 6	2	3. 36. 16, 6✠	2
8	.	13. 54. 37, 4:::	1	8. 35. 59, 0✠	1
5	♓ e	13. 57. 25, 0	2	4. 21. 14, 5✠	2
8	.	14. 32. 10, 0	2	8. 14. 38, 6✠	2
7.8	.	14. 41. 0, 7	1	8. 59. 22, 8✠	1
6	♓ 1 ζ	15. 15. 16, 5	2	6. 16. 46, 6✠	2
7	♓ 2 ζ	15. 15. 38, 9	2	6. 16. 59, 5✠	2
7	..	15. 38. 11, 1	1	5. 49. 4, 8✠	1
6	♓ f	16. 18. 36, 6	2	2. 19. 26, 8✠	2
8.9	.	16. 42. 4, 0	1	2. 0. 11, 0✠	1
6	.	16. 50. 24, 0	1	1. 47. 45, 8—	1
7	.	17. 31. 15, 7	1	0. 26. 48, 8✠	1

G 2

Magn.	characteres	Ascens. recta.	Num. Obseru.	Declinatio.	Num. Obseru.
8	.	17. 31. 48, 0	1	3. 27. 20, 3✠	1
7.8	.	17. 33. 24, 0	1	6. 7. 17, 5✠	1
5	♓ 1 ρ	18. 17. 8, 5	3	17. 53. 44, 6✠	3
5	♓ 2 ρ	18. 23. 17, 8	3	17 58. 8, 5✠	3
7	..	18. 49. 44, 6	1	15. 48. 32, 5✠	1
5	♓ μ	19. 21. 27, 1	3	4. 52. 47, 0✠	3
4.3	♓ η	19. 36. 56, 8	3	14. 4. 51, 9✠	3
7.8	.	19. 42. 45, 5	1	9. 37. 27, 7✠	1
7.8	.	20. 8. 12, 2	2	6. 56. 59, 4✠	2
8	..	20. 31. 46, 1	1	7. 1. 6, 0✠	1
5.4	♓ π	21. 2. 58, 6	4	10. 53. 9, 3✠	4
8.9	.	21. 10. 57, 7	3	10. 49. 35, 5✠	3
7.8	.	21. 56. 30, 4	2	7. 30. 56, 2✠	2
9	.	22. 9. 6, 7	1	7. 49. 49, 5✠	1
5	♓ ν	22. 11. 27, 7	4	4. 14. 43, 9✠	4
5	♓ ο	23. 8. 10, 2	2	7. 55. 18, 7✠	2
8.9	.	23. 33. 33, 1	1	9. 36. 54, 6✠	1
7.8	.	24. 29. 19, 1	3	9. 49. 35, 3✠	3
4	♈ 1 γ	25. 2. 46, 6	4	18. 5. 24, 3✠	4
.	♈ 2 γ	25. 2. 48, 5	4	18. 5. 15, 2✠	4
3	♈ β	25. 18. 4, 1	2	19. 36. 20, 5✠	2
5	♈ ι	26. 0. 51, 8	4	16. 37. 0, 0✠	3
8	.	26. 36. 3, 0	2	11. 6. 4, 7✠	2
2	Andromedae γ	27. 15. 14, 1	4	41. 8. 48, 1✠	4
3	♓ α	27. 21. 50, 8	2	1. 34. 34, 8✠	2
6	♈ κ	28. 14. 23, 0	2	21. 28. 27, 9✠	2
2	lucida ♈ α	28. 22. 1, 5	4	22. 17. 50, 9✠	4
8	.	28. 36. 10, 8	2	16. 51. 26, 7✠	2
7	.	29. 17. 6, 6	1	18. 20. 17, 5✠	1
9	.	29. 26. 5, 2	1	20. 12. 55, 3✠	1
6	♈ η	29. 47. 39, 0	2	20. 3. 10, 0✠	2
4	Ceti 1 ξ	30. 1. 37, 1	2	7. 41. 32, 4✠	2
7	.	30. 56. 43, 5	1	14. 7. 30, 2✠	1
6	♈ θ	31. 8. 52, 4	2	18. 45. 37, 4✠	2
7	.	31. 51. 32, 7	2	5. 28. 36, 2—	2
5	♈ ξ	32. 56. 39, 8	2	9. 29. 36, 4✠	2
4	Ceti 2 ξ	33. 48. 23, 8	2	7. 21. 16, 5✠	2
7.8	.	34. 14. 57, 1	1	18. 45. 33, 3✠	1
4.5	Ceti ν	35. 46. 33, 5	1	4. 30. 59, 5✠	1
6	♈ ν	36. 15. 5, 5	1	20. 53. 33, 9✠	1
.	..	36. 45. 9, 6	1	0. 44. 7, 8—	1
6	♈ μ	37. 9. 42, 9	1	18. 57. 30, 5✠	1
6	♈ ο	37. 47. 4, 1	2	14. 15. 54, 5✠	2
4	Ceti μ	37. 56. 50, 6	2	9. 4. 14, 0✠	4
6	♈ π	38. 55. 41, 0	3	16. 26. 7, 2✠	3

Magn.	Stellarum characteres.	Afcenf. recta	Num. Obferu.	Declinatio.	Num. Obferu.
6	♈ σ	39. 30. 50, 9	2	14. 3. 44, 1 +	2
6	♈ 1 ρ	40. 32. 3, 9	1	17. 19. 51, 1 +	1
.	♈ 2 ρ	40. 40. 32, 7	1	17. 2. 9, 2 +	1
5	♈ ε	41. 19. 31, 7	3	20. 20. 57, 4 +	3
4	Ceti λ	41. 40. 6, 4	2	7. 55. 13, 8 +	2
2	Mandib. Ceti α	42. 23. 23, 5	2	3. 7. 9, 4 +	2
3	Medufae caput	43. 5. 37, 0	3	39. 59. 50, 0 +	4
8	.	43. 50. 48, 0:	1	19. 48. 55, 2:: +	1
7	.	43. 55. 59, 3	1	18. 25. 56, 4 +	1
4	♈ δ	44. 25. 44, 4	3	18. 47. 13, 1 +	3
5	♈ ζ	45. 13. 44, 5	3	20. 7. 27, 1 +	3
2	Perfei α	46. 45. 40, 2	3	48. 58. 12, 9 +	3
5	♈ 1 τ	46. 47. 38, 8	2	20. 15. 2, 4 +	3
6	♈ 2 τ	47. 11. 13, 4	3	19. 51. 7, 8 +	3
7	.	47. 36. 7, 0	1	19. 55. 8, 2 +	1
4	♉ o	47. 55. 48, 0	4	8. 9. 13, 6 +	4
4	♉ ξ	48. 29. 44, 3	3	8. 51. 54, 0 +	3
7	.	48. 49. 40, 5	1	10. 31. 48, 9 +	1
7.8	.	48. 59. 37, 7	1	15. 54. 14, 1 +	1
6	♉ s	49. 16. 37, 2	3	10. 28. 54, 5 +	3
5	♉ f	49. 21. 33, 4	3	12. 5. 1, 4 +	4
6	♉ t	49. 51. 26, 9	3	8. 31. 55, 1 +	3
7	.	50. 4. 38, 8:::	1	17. 0. 57, 0 +	1
7	.	51. 2. 0, 7	1	14. 36. 39, 1 +	1
6	Pleiadum Celeno	52. 35. 13, 3	2	23. 30. 9, 4 +	1
5	Electra	52. 36. 28, 9	5	23. 19. 35, 8 +	4
7	.	52. 39. 38, 7	1	23. 51. . . . +	1
5	Taigeta	52. 40. 56, 9	4	23. 40. 55, 3 +	3
6	Maia	52. 50. 17, 8	4	23. 35. 9, 0 +	3
7	Afterope 1	52. 51. 14, 7	1	23. 46. 25, 5 +	1
7	Afterope 2	52. 53. 23, 7	1	23. 44. 48, 5 +	1
5	Merope	52. 58. 23, 7	1	23. 10. 10, 8 +	1
7	p	53. 13. 20, 7	1	23. 20. 33, 3 +	1
3	♉ η, Alcyone	53. 15. 21, 3	5	23. 19. 51, 7 +	5
5	Atlas	53. 40. 28, 0	3	23. 17. 13, 9 +	3
6	Pleione	53. 40. 40, 0	3	23. 22. 17, 2 +	3
7.8	.	53. 42. 58, 3	1	23. 36. 57, 0 +	1
5	♉ e	53. 43. 49, 6	1	10. 22. 28, 5 +	1
7	.	54. 48. 43, 1	1	16. 34. 52, 1 +	1
.	.	55. 37. 20, 3	1	21. 45. 19, 8 +	1
4	♉ λ	56. 47. 56, 8	4	11. 46. 54, 6 +	3
5	♉ 1 A	57. 34. 38, 4	3	21. 23. 40, 2 +	4
7	♉ 2 A	57. 43. 55, 8	2	21. 19. 47, 9 +	3
5	♉ ψ	57. 59. 25, 6	1	28. 19. 12, 0 +	2
6	♉ 1 ω	58. 44. 38, 2	2	18. 56. 37, 2 +	2

Magn.	Stellarum characteres	Afcenf. recta	Num. Obferu.	Declinatio.	Num. Obferu.
6	♉ p	59. 0. 11, 1	2	25. 49. 21, 1 +	2
6	♉ 2 ω	60. 44. 53, 6	4	19. 57. 25, 3 +	4
5	♉ φ	61. 20. 49, 6	4	26. 44. 41, 2 +	4
3	♉ γ	61. 29. 1, 1	5	15. 1. 3, 9 +	5
6.7	♉ 1 h	61. 33. 35, 3::	1	13. 25. 34, 6 +	1
.	♉ 2 h	61. 42. 0, 9	1	14. 29. 26, 3 +	1
.	.	61. 47. 46, 1	1	13. 15. 35, 7 +	1
5	♉ χ	61. 56. 29, 6	2	25. 1. 54, 5 +	2
.	♉ 3 h	62. 5. 0, 8	1	13. 28. 45, 9 +	1
4	♉ 1 δ	62. 13. 23, 7	4	16. 56. 56, 9 +	5
.	.	62. 21. 30:::	.	16. 0 . . . +	.
5	♉ 2 δ	62. 30. 53, 9	5	16. 51. 27, 0 +	5
6	♉ 3 δ	62. 51. 2, 1	4	17. 20. 53, 7 +	5
5	♉ 1 υ	62. 56. 0, 8	1	22. 14. 21, 9 +	2
.	f	63. 6. 57, 5	1	15. 2. 39, 2 +	1
6	♉ 2 υ	63. 10. 59, 1	2	22. 25. 29, 9 +	2
5	♉ π	63. 12. 37, 5	2	14. 8. 27, 0 +	3
7	.	63. 12. 50:::	.	16. 0. . . . +	.
3.4	♉ ε	63. 35. 57, 1	4	18. 37. 3, 6 +	5
.	.	63. 38. 55, 9	1	14. 10. 36, 9 +	1
5	♉ 1 θ	63. 39. 57, 4	2	15. 24. 0, 6 +	2
5	♉ 2 θ	63. 41. 22, 9	2	15. 18. 32, 3 +	2
8	.	64. 0. 15, 8	1	19. 17. 11, 8 +	1
.	.	64. 3. 46, 4	1	15. 10. . . . +	1
.	.	64. 9. 20, 4	4	15. 38. 31, 2 +	5
.	.	64. 11. 22, 4	1	15. 8. 27, 5 +	1
7	.	64. 12. 38, 7	2	15. 35. 50, 3 +	2
7.8	.	64. 55. 31, 5	3	15. 47. 14, 9 +	3
5	♉ ρ	65. 0. 29, 3	1	14. 18. 38, 3 +	1
8	.	65. 15. 0:::	.	16. 0 . . . +	.
.	.	65. 21. 35, 7	1	19. 21. 18, 9 +	1
1	Aldebaran ♉ α	65. 29. 12, 0	12	15. 59. 48, 1 +	18
.	♉ p	66. 3. 14, 7	4	15. 31. 22, 2 +	4
7	♉ 1 σ	66. 18. 46, 6	5	15. 17. 49, 6 +	7
6	♉ 2 σ	66. 20. 11, 5	5	15. 24. 46, 3 +	7
5	♉ τ	66. 54. 27, 4	5	22. 27. 59, 7 +	6
7	.	68. 0. 50, 9	6	18. 16. 15, 2 +	6
8	.	68. 36. 34, 1	3	18. 16. 48, 1 +	3
6	i	69. 16. 51, 1	6	18. 24. 7, 3 +	6
4.5	Orionis o	69. 41. 18, 2	3	13. 49. 21, 1 +	3
9	.	69. 59. 33, 2	1	16. 36. 7, 1 +	1
8	.	70. 13. 47, 6	1	15. 58. 1, 6 +	1
8	.	70. 19. 38, 3	1	16. 12. 15, 0 +	1
6.7	.	70. 45. 30, 8	3	23. 32. 38, 6 +	3
6	.	70. 49. 29, 6	3	16. 44. 53, 3 +	4

Magn.	Stellarum characteres	Ascens. recta.	Num. Obseru.	Declinatio.	Num. Obseru.
6.7	.	71. 17. 30,7	1	14. 8. 50,8 +	1
7	.	71. 26. 22,3	1	15. 31. 48,7 +	1
4	♉ ι	72. 8. 1,8	5	21. 13. 3,6 +	5
.	.	72. 27. 39,0	3	20. 54. 44,8 +	3
.	.	72. 39. 41,8	2	15. 2. 35,1 +	1
.	.	72. 47. 34,2	3	19. 26. 51,7 +	3
6	♉ m	73. 15. 41,0	4	18. 17. 38,2 +	7
6	♉ l	73. 20. 50,5	2	20. 4. 24,8 +	2
6	.	73. 56. 16,7	2	15. 15. 48,0 +	3
1	Capella, Aur. α	74. 40. 34,9	13	45. 43. 10,7 +	25
7	.	75. 12. 10,5	1	21. 58. 55,4 +	1
1	Rigel, Orionis β	75. 42. 23,4	....	8. 30. 9,5 —	14
6	♉ n	76. 9. 31,2	2	21. 49. 13,7 +	3
7	.	76. 12. 16,2	2	19. 51. 18,2 +	2
8	.	76. 28. 52,5	2	18. 14/18. 46/17, 9/8 +	2
7	.	76. 38. 2,4	2	19. 32. 35,7 +	3
.	.	76. 41. 47,7:::	1	27. 41. 4,7 +	1
7.8	.	76. 59. 12,5	2	28. 40. 44,9 +	2
2	♉ β	77. 43. 17,7	6	28. 22. 32,3 +	10
2	Orionis γ	78. 0. 58,5	...	6. 6. 25,2 +	7
.	.	78. 14. 11,7	3	17. 43. 39,6 +	2
5	o	78. 14. 54,0	3	21. 42. 15,5 +	3
5	♉ χ	79. 12. 57,7	3	31. 59. 6,0 +	3
5	Leporis β	79. 26. 59,1	...	20. 58. 15,6 —	5
7.8	.	79. 47. 18,8	2	20. 16. 39,2 +	2
2	Orionis δ	79. 53. 23,1	...	0. 30. 3,0 —	11
6	.	80. 1. 10,0	1	14. 8. . . . +	1
3	Leporis α	80. 29. 40,8	..	18. 0. 53,3 —	4
3	♉ ζ	80. 46. 6,2	3	20. 58. 10,9 +	4
2	Orionis ε	80. 57. 43,7	..	1. 22. 40,2 —	9
.	.	81. 9. 28,7	1	25. 44. 7,5 +	1
4	Orionis σ	81. 37. 42,8	...	2. 45. 38,8 —	8
2	Orionis ζ	82. 6. 55,0	..	2. 5. 30,2 —	11
.	.	82. 43. 5,0	2	18. 34. 31,3 +	2
7.8	.	83. 6. 53,4	1	20. 9. 47,9 +	1
7.8	.	83. 14. 21,0	1	24. 33. 54,1 +	1
.	.	83. 18. 12,6	2	17. 36. 50,1 +	2
.	.	83. 27. 23,6.·.	1	20. 45. 40,3 +	1
6.7	.	83. 30. 46,4	1	24. 27. 35,7 +	1
4.3	Leporis γ	83. 34. 31,2	...	22. 32. 40,0 —	4
8	.	83. 37. 30,5	1	9. 38. 44,9 +	1
3	Orionis κ	84. 2. 53,5	..	9. 46. 31,4 —	5
7.8	.	84. 42. 30,5	1	20. 13. 3,0 +	1
5	Orionis 1 χ	84. 59. 4,0	3	20. 11. 26,7 +	4
7	.	85. 7. 30,3	1	19. 40. 38,5 +	1

Magn.	Stellarum characteres.	Ascens. recta	Num. Obseru.	Declinatio.	Num. Obseru.
1	Orionis α	85. 29. 36,9	...	7. 20. 18,9 +	8
.	.	85. 42. 52,7	3	25. 53. 52,3 +	3
8	.	86. 25. 10,1	1	22. 51. 40,8 +	1
7	.	86. 44. 26,5	1	22. 22. 8,3 +	1
5	Orionis 2 χ	87. 15. 8,5	1	19. 40. 15,0 +	1
6.7	.	87. 15. 19,7	1	17. 47. 35,6 +	1
4	♊ H	87. 19. 25,1	3	23. 15. 6,8 +	3
7	Orionis 3 χ	87. 21. 28,1	2	20. 7. 12,7 +	3
.	.	88. 15. 17,0	1	22. 30. . . . +	1
7.8.	.	88. 42. 32,8	1	22. 12. 14,9 +	1
.	.	88. 43. 49,6	2	23. 7. 43,6 +	2
7.8.	.	88. 55. 42,7	1	23. 1. 13,5 +	1
6	.	89. 8. 36,1	1	24. 26. 57,3 +	1
7.8	.	89. 22. 56,2	3	22. 56. 21,5 +	3
6	.	89. 23. 34,1	1	19. 49. 11,1 +	1
4	♊ η	90. 2. 13,3	4	22. 33. 11,5 +	5
6	.	90. 7. 17,4	1	19. 12. 53,7 +	1
.	.	90. 21. 7,3	1	24. 1. 28,2 +	1
.	.	90. 31. 20,0	2	23. 45. 48,2 +	2
.	.	91. 0. 35,0	2	23. 40. 24,3 +	2
4	.	91. 7. 2,9	1	23. 32. 26,2 +	1
.	.	91. 8. 7,2	1	23. 20. 44,6 +	1
8	.	91. 39. 42,0	1	21. 13. 1,1 +	1
9	.	91. 41. 34,6	1	21. 17. 10,3 +	1
.	.	91. 43. 46,4	1	23. 50. 59,4 +	1
3	♊ μ	92. 2. 51,0	3	22. 36. 53,6 +	3
.	.	92. 46. 0,7	2	21. 45. 19,7 +	2
.	.	93. 18. 27,1	3	20. 54. 58,6 +	1
.	.	93. 21. 25,8	3	20. 37. 6,7 +	3
4	♊ ν	93. 37. 3,2	4	20. 20. 35,4 +	3
8	.	94. 30. 40,8	3	17. 55. 40,8 +	2
7.8	.	94. 30. 57,1	2	17. 55. 57,2 +	2
8	.	95. 12. 24,3	2	16. 22. 17,8 +	2
2.3	♊ γ	95. 54. 7,7	5	16. 35. 4,5 +	5
8	.	96. 24. 10,5	2	19. 51. 30,5 +	2
.	.	96. . . .	2	16. 36. 5,9 +	2
6.7	.	97. 2. 46,4	5	17. 51. 38,4 +	6
3	♊ ε	97. 13. 37,3	5	25. 20. 51,6 +	5
5.4	♊ 1 ξ	97. 33. 23,4	3	13. 27. 6,9 +	4
.	♊ 2 ξ	97. 53. 53,1	5	13. 8. 12,0 +	5
1	Sirius.	98. 35. 53,0	...	16. 23. 54,2 —	18
.	.	99. 0 . . .	.	13. 40. 19,4 +	1
6	♊ d	99. 13. 40,8	3	22. 1. 29,2 +	3
6	♊ e	100. 13. 5,8	5	13. 27. 54,3 +	5
7.8	.	100. 31. 6,2	3	18. 1. 50,4 +	3

Magn.	Stellarum characteres.	Ascens. recta	Num. Observ.	Declinatio.	Num. Observ.
	.	100. 52. 25, 8	2	18. 11. 59, 7 +	2
7.8	.	100. 55. 56, 8	1	26. 22. 43, 2 +	1
7	.	101. 5. 41, 5	1	26. 13. 15, 1 +	1
9	.	101. 27. 55, 0	1	23. 45. 21, 9 +	1
8.9	.	101. 33. 24, 3	1	16. 23. 37, 3 +	1
6	♊ 1ω	101. 52. 55, 2	5	24. 32. 22, 2 +	7
3	♊ ζ	102. 24. 18, 0	6	20. 54. 18, 2 +	8
.	♊ 2ω	102. 39. 0, 0	5	22. 58. 46, 8 +	6
8	.	103. 1. 28, 2	1	18. 5. 31, 2 +	1
7	.	103. 35. 24, 5	1	16. 17. 55, 1 +	1
5	♊ τ	103. 53. 43, 8	5	30. 37. 14, 1 +	6
6	♊ m	104. 23. 48, 4	4	24. 30. 45, 6 +	6
5	♊ 1λ	104. 50. 12, 1	4	16. 33. 2, 6 +	6
5	♊ n	104. 56. 18, 1	4	25. 17. 5, 3 +	6
7.8	.	105. 29. 12, 5	1	27. 6. 17, 9 +	1
7	.	105. 40. 56, 6	1	16. 33. 10, 6 +	1
5	♊ 2λ	106. 0. 50, 4	6	16. 57. 30, 1 +	9
3	♊ δ	106. 22. 54, 8	4	22. 24. 30, 3 +	7
6.7	♊ q	106. 52. 56, 1	4	20. 32. 51, 5 +	7
5.6	♊ A	107. 8. 40, 0	4	25. 29. 45, 9 +	7
7	.	107. 20. 11, 5	1	28. 5. 3, 8 +	1
7.8	.	107. 20. 56, 0	1	18. 43. 16, 2 +	1
4.5	♊ ι	107. 38. 9, 4	4	28. 15. 32, 6 +	6
7.8	.	108. 5. 49, 7	1	22. 0. 2, 5 +	1
6	♊ p	108. 18. 28, 7	4	21. 55. 17, 0 +	7
5	♊ ρ	108. 20. 40, 9	4	32. 14. 40, 1 +	7
6	♊ 1b	108. 31. 38, 9	3	28. 35. 49, 6 +	6
.	♊ 2b	108. 39. 4, 8	3	28. 23. 45, 4 +	7
1.2	Castor ♊ α	109. 44. 44, 8	...	32. 23. 53, 1 +	15
6	♊ k	109. 54. 56, 1	4	16. 19. 46, 2 +	5
5	♊ υ	110. 12. 46, 5	5	27. 24. 52, 5 +	5
.	.	110. 49. 3, 6	1	19. 26. 47, 1 +	1
7	.	110. 51. 38, 6	1	24. 53. 0, 3 +	1
7	.	111. 6. 23, 6	1	24. 45. 10, 8 +	1
6	♊ f	111. 20. 26, 0	2	18. 12. 33, 7 +	3
1	Procyon	111. 37. 48, 0	...	5. 49. 51, 3 +	22
7.8	. .	111. 47. 27, 6	1	5. 49. . . . +	.
5	♊ σ	112. 0. 23, 2	3	29. 27. 3, 1 +	3
7	.	112. 12. 5, 0	1	22. 57. 5, 4 +	1
6	♊ c	112. 18. 0, 0	7	26. 20. 34, 1 +	8
4.5	♊ κ	112. 25. 16, 8	2	24. 57. 33, 8 +	2
2	Pollux ♊ β	112. 35. 12, 7	...	28. 35. 36, 6 +	13
6	♊ g	112. 59. 34, 2	2	19. 5. 3, 5 +	3
.	.	114. 26. 0, 0	1	19. 55. 41, 0 +	1
6	♊ φ	114. 37. 55, 5	2	27. 22. 28, 1 +	2

Magn.	Stellarum characteres	Ascens. recta.	Num. Observ.	Declinatio.	Num. Observ.
7	♊ l	115. 20. 56, 2	3	20. 30. 25, 7 +	3
6	♋ ω	116. 31. 56, 9	2	26. 2. 18, 3 +	4
5	♊ χ	117. 7. 21, 6	3	28. 27. 22, 2 +	5
.	..	117. 51. 56, 2	3	13. 47. 37, 5 +	4
7	.	117. 58. 30, 7	1	14. 10. 42, 7 +	1
5	♋ μ	118. 20. 33, 1	3	22. 16. 9, 5 +	5
7	.	118. 51. 16, 0	1	26. 32. 28, 7 +	1
4	♋ ψ	118. 55. 39, 8	2	26. 13. 36, 3 +	3
.	.	119. 0. 52, 4	1	17. 42. 49, 7 +	1
6	♊ ψ	119. 29. 38, 8	3	30. 21. 59, 2 +	3
5	duplex ♋ ζ [.	119. 32. 50, 8	4	18. 21. 45, 3 +	7
	.	119. 32. 52, 3	4	18. 21. 48, 3 +	7
7	.	119. 44. 16, 6	1	10. 31. 35, 6 +	1
.	.	120. 2. 52, 0	1	18. 20. 0, 1 +	1
4.3	♋ β	120. 49. 0, 1	4	9. 55. 7, 9 +	7
.	.	121. 4. 27, 1	1	9. 36. 9, 6 +	1
6	♋ χ	121. 17. 55, 7	4	27. 59. 14, 3 +	7
6	♋ λ	121. 29. 48, 4	6	24. 46. 15, 5 +	7
7	.	122. 20. 23, 2	6	19. 5. 47, 9 +	6
6.7	♋ 1φ	122. 53. 36, 2	4	28. 40. 34, 4 +	5
7	.	122. 59. 46, 5	2	17. 49. 50, 9 +	2
.	♋ 2φ	122. 59. 47, 2	4	27. 42. 44, 0 +	5
7	.	123. 1. 35, 7	1	24. 57. 57, 0 +	1
6.7	.	123. 31. 33, 8	5	24. 56. 5, 7 +	6
8	.	123. 46. 55, 0	2	26. 59. 9, 6 +	2
7	.	123. 47. 9, 1	1	25. 8. 20, 1 +	1
6	♋ υ	124. 15. 35, 8	3	24. 53. 4, 2 +	4
6.5	♋ θ	124. 24. 40, 9	5	18. 54. 4, 1 +	6
.	.	124. 24. 48, 0	2	19. 47. 30, 0 +	2
.	.	124. 38. 6, 8	1	24. 53. 48, 6 +	1
6.7	♋ η	124. 38. 26, 0	5	21. 15. 4, 5 +	4
.	.	125. 3. 6, 8	1	14. 4. 25, 1 +	2
8	.	125. 18. 53, 4	4	20. 24. 38, 3 +	5
8	..	125. 25. 5, 2	2	20. 35. 38, 2 +	2
8	.	125. 36. 11, 0	2	16. 8. 25, 2 +	2
.	.	125. 56. 20, 6	6	20. 5. 56, 5 +	6
6	♋ c	125. 57. 41, 4	3	10. 29. 7, 8 +	2
.	.	126. 12. 47, 1	4	10. 24. 36, 8 +	4
7	.	126. 24. 52, 9	4	20. 37. 8, 1 +	3
7	.	126. 27. 10, 0	3	20. 22. 57, 3 +	2
7	.	126. 30. 30, 2	2	20. 51. 5, 6 +	1
.	.	126. 30. . . .	.	21. 15. 13, 1 +	1
7	.	126. 31. 49, 0	2	20. 48. 53, 7 +	1
7	.	126. 34. 53, 5	3	20. 30. 52, 5 +	3
7	.	126. 36. 19, 5	3	20. 23. 21, 7 +	2

Magn.	characteres.	Ascens. recta	Observ.	Declinatio.	Num. Observ.
7	.	126. 40. 8, 7	2	20. 33. 56, 2 +	2
7	.	126. 43. 40, 9	2	20. 25. 38, 4 +	2
7	.	126. 56. 45, 5	3	20. 43. 33, 4 +	4
4	♋ γ	127. 16. 55, 7	6	22. 19. 40, 4 +	6
6	♋ 1 A	127. 26. 2, 8	4	13. 32. 19, 2 +	4
4	♋ δ	127. 41. 45, 6	9	19. 2. 2, 0 +	9
6	♋ b	127. 52. 18, 7	5	10. 56. 54, 3 +	7
5	♋ ι	127. 58. 13, 6	2	29. 38. 1, 8 +	2
6	♋ 2 A	128. 22. 59, 8	2	12. 59. 15, 2 +	4
8	.	128. 48. 17, 6	2	13. 25. 42, 4 +	2
8	.	129. 12. 6, 8	1	18. 5^{7}_{3}. $^{11}_{41}$, $^{8}_{0}$ +	2
7.8	.	129. 12. 21, 1	1	19. 43. 27, 6 +	1
8	.	129. 21. 4, 2	4	16. 14. 20, 5 +	6
7.8	.	129. 51. 12, 8	3	18. 16. 26, 8 +	5
.	.	129. 58. 48, 2	2	20. 52. 22, 9 +	2
9	.	130. 0. 32, 1	1	18. 27. 4, 7 +	1
.	.	130. 2. 32, 3	1	15. 18. 57, 3 +	1
9	.	130. 13. 15, 9:	1	15. 10. 50, 7::: +	1
9	.	130. 20. 24, 7:	1	9. 9. 38, 1: +	1
7	.	130. 23. 59, 5	3	18. 8. 36, 5 +	3
8.9	.	130. 31. 2, 4	1	15. 5. 57, 8: +	1
6.7	.	130. 38. 38, 1	3	12. 32. 30, 8 +	3
.	.	130. 50. 43, 9	2	18. 3. 58, 8 +	2
6	♋ 1 ο	130. 54. 4, 8	5	16. 14. 32, 1 +	8
.	.	130. 56. . . .	.	18. 22. 43, 1 +	1
6	♋ 2 ο	130. 58. 56, 1	5	16. 30. 5, 2 +	6
4.3	♋ α	131. 16. 45, 2	8	12. 47. 11, 8 +	10
7	.	131. 20. 24, 9	1	19. 4. 5, 0 +	1
.	.	131. 30. 6, 4	2	14. 0. 27, 6 +	2
6	♋ ν	132. 6. 21, 8	8	25. 23. 43, 7 +	7
7.8	.	132. 50. 3, 9	3	11. 48. 14, 7 +	2
4.5	♋ κ	133. 37. 36, 3	6	11. 38. 4, 0 +	5
5.6	♋ 1 ξ	133. 49. 20, 5	5	23. 1. 0, 4 +	5
.	♋ 2 ξ	134. 4. 19, 3	5	22. 58. 14, 8 +	5
..	.	134. 30. . . .	.	12. 32. 39, 7 +	1
7	.	134. 43. 39, 7	3	15. 57. 52, 6 +	3
7	♋ π	135. 25. 54, 0	6	15. 56. 19, 0 +	6
7	.	136. 1. . . .	2	12. 30. 35, 0 +	2
7	.	136. 13. 57, 4	1	27. 22. 51, 0 +	1
7.8	.	136. 35. 27, 9:	1	10. 48. 16, 1 +	1
8.9	.	136. 42. 53, 1	1	20. 6. 33, 7 +	1
7	.	137. 28. 16, 4	1	26. 57. 8, 1 +	2
4	♌ κ	137. 35. 50, 5	5	27. 13. 2, 3 +	6
7.8	.	137. 37. 12, 3:	1	8. 44. 51, 5 +	1
5	♌ ω	138. 50. 32, 0	6	10. 6. 20, 8 +	5

Magn.	characteres.	Ascens. recta	Observ.	Declinatio.	Num. Observ.
2	Hydrae cor	138. 53. 59, 2	3	7. 36. 46, 1 —	4
4	♌ λ	139. 26. 14, 1	7	24. 1. 46, 1 +	7
4	♌ ξ	139. 41. 31, 1	7	12. 22. 1, 0 +	6
6	♌ h	139. 42. 50, 9	5	10. 46. 43, 2 +	5
7.8	.	140. 25. 45, 2	2	13. 43. 46, 2 +	2
6.7	.	140. 53. 8, 4	1	17. 31.	1
.	.	141. 4. 41, 2	6	7. 55. 2, 7 +	8
.	.	141. 30. 5, 7	1	14. 24. 5, 5 +	1
5.4	♌ ο	142. 1. 34, 3	7	10. 59. 24, 1 +	10
6	♌ ψ	142. 36. 9, 7	9	15. 7. 33, 4 +	11
3	♌ ε	142. 59. 13, 0	6	24. 53. 5, 3 +	8
6.7	.	143. 18. 10, 9	3	12. 55. 19, 2 +	3
8	.	143. 29. 50, 6	1	21. 43. 3, 1 +	1
7.8	.	143. 34. 15, 0	3	12. 41. 6, 8 +	3
6	.	143. 36. 50:::	1	12. 33. 4, 1 +	1
8	.	144. 24. 58, 9	2	12. 58. 18, 7 +	2
7	.	144. 26. 59, 6	5	14. 11. 43, 7 +	5
7.8	.	145. 17. 39, 9	1	9. 12. 53, 0 +	1
6.7	.	145. 52. 10, 1	2	10. 4. 42, 1 +	2
4.5	♌ υ	146. 16. 1, 2	9	13. 35. 50, 0 +	11
7	.	146. 17. 41, 1	2	9. 28. 5, 5 +	2
4	♌ π	146. 49. 28, 8	10	9. 12. 14, 5 +	12
8	.	147. 38. 16, 6	3	11. 5. 5, 1 +	3
8	.	147. 45. 38, 1	2	12. $^{47}_{51}$. $^{55}_{26}$, $^{5}_{4}$ +	2
8.9	.	148. 2. 28, 7	2	9. 9. 54, 9 +	2
7.8	.	148. 6. 1, 7	2	16. 55. 51, 0 +	2
3.4	♌ η	148. 29. 52, 9	9	17. 56. 31, 0 +	12
5	♌ A	148. 44. 0, 1	6	11. 10. 59, 6 +	7
1	Regulus ♌ α	148. 50. 15, 7	. . .	13. 9. 0, 0 +	26
8	.	149. 8. 21, 2	1	10. 46. 46, 0 +	1
7.8	.	149. 37. 9, 5	2	14. 33. 1, 7 +	2
8	.	150. 15. 20, 6	4	22. 22. 23, 4 +	4
8	.	150. 18. 55, 8	1	17. 20. 21, 9 +	1
.	.	150. 43. 58, 5	3	18. 56. 42, 1 +	3
6.7	.	150. 53. 15, 9	3	14. 56. 10, 2 +	3
7.8	.	151. 20. 19, 0	1	13. 50. 5, 3 +	1
6	♌ 1 γ	151. 36. 9, 4	8	20. 41. 58, 7 +	10
2	♌ 2 γ	151. 37. 8, 4	8	21. 3. 59, 8 +	10
7	.	152. 10. 11, 3	2	16. 11. 52, 0 +	3
7.8	.	152. 20. 17, 3	3	10. 11. 12, 4 +	4
8	.	152. 50. 40, 5	2	11. 48. 54, 7 +	3
7	.	153. 5. 42, 7	5	10. 1. 0, 3 +	4
7	♌ 1 ρ	153. 41. 6, 2	4	10. 59. 54, 0 +	5
6	.	153. 55. 55, 1	1	15. 34. 52, 8 +	1
7.8	.	154. 4. 8, 0	2	11. 23. 47, 9 +	2

Magn.	Stellarum characteres	Ascens. recta	Num. Obseru.	Declinatio.	Num. Obseru.
6	♌ i	154. 47. 6, 9	4	15. 22. 54, 4 ✠	5
4	♌ 2ρ	154. 59. 12, 6	4	10. 33. 15, 9 ✠	6
7	.	155. 30. 50, 9	4	8. 12. 10, 5 ✠	4
7	..	155. 33. 15, 2	3	9. 54. 11, 7 ✠	3
9	..	156. 9. 34, 3	1	18. 33. 24, 8: ✠	1
7	.	156. 26. 48, 9	1	17. 23. 25, 1 ✠	1
9	.	156. 43. 27, 0	2	10. 6. 22, 4 ✠	2
6	.	157. 40. 11, 8	1	6. 1. 16, 2 ✠	1
.	.	158. 12. 4, 6	1	8. 47. 31, 8 ✠	1
7.8	.	158. 21. 15, 1::	1	7. 39. 12, 1 ✠	1
6	♌ k	158. 22. 5, 4	2	15. 28. 32, 0 ✠	1
8	.	158. 39. 18, 0	1	7. 37. 39, 8 ✠	1
6	♌ l	159. 6. 10, 2	1	11. 49. 49, 5 ✠	1
8.9.	.	160. 31. 22, 2	1	7. 8. 27, 0 ✠	1
5	♌ d	161. 59. 23, 7	2	4. 55. 21, 2 ✠	1
5	♌ c	162. 1. 24, 4	1	7. 24. 28, 2 ✠	1
.	.	162. 50. 56, 3	1	4. 56. 56, 0 ✠	1
4	♌ x	163. 6. 19, 9	3	8. 38. 57, 7 ✠	3
7.8	.	165. 12. 23, 1	1	21. 27. 38, 5 ✠	1
2.3	♌ δ	165. 16. 21, 6	3	21. 51. 22, 1 ✠	3
3	♌ θ	165. 21. 8, 8	3	16. 45. 34, 1 ✠	3
6	♌ η	165. 46. 7, 7	3	14. 38. 4, 7 ✠	3
4	♌ φ	166. 4. 7, 4	2	2. 19. 14, 9 —	2
4.5	♌ σ	167. 8. 21, 4	2	7. 21. 48, 5 ✠	2
4	♌ ι	167. 47. 55, 1	2	11. 52. 11, 6 ✠	2
.	.	167. 52. 56, 5	2	2. 44. 41, 2 ✠	2
7	.	168. 36. 13, 1	1	4. 20. 27, 4 ✠	1
4	♌ τ	168. 50. 57, 7	2	4. 11. 52, 3 ✠	3
4.5	♌ e	169. 27. 57, 1	2	1. 39. 33, 5 —	3
7.8	.	169. 53. 20, 4	1	5. 7. 9, 7 —	1
7	. .	170. 6. 7, 4	1	6. 28. 57, 3 —	2
4	♌ υ	171. 7. 1, 8	2	0. 31. 17, 3 ✠	2
6	♍ ω	171. 28. 6, 6	2	9. 29. 0, 6 ✠	2
8	.	172. 24. 48, 2	1	6. 5. 49, 6: ✠	1
5	♍ ξ	173. 10. 31, 8	2	9. 36. 46, 4 ✠	2
5	♍ ν	173. 19. 49, 3	2	7. 53. 43, 2 ✠	2
.	♍ ·1 A	173. 50. 40, 0	2	9. 35. 59, 3 ✠	2
1.2	Cauda ♌ β	174. 8. 57, 8	3	15. 56. 6, 1 ✠	4
3	♍ β	174. 29. 58, 8	2	3. 8. 15, 7 ✠	3
6	♍ 2 A	175. 37. 50, 2	1	9. 48. 0, 2 ✠	2
8	.	175. 43. 50, 0	1	2. 27. 21, 1 ✠	1
7	.	176. 26. 2, 0	1	4. 50. 30, 1 ✠	1
.	.	176. 38. 51, 3	1	1. 53. 12, 7 ✠	1
5.6	♍ b	176. 51. 54, 3	3	5. 0. 52, 8 ✠	4
5	♍ π	177. 5. 32, 2	3	7. 58. 30, 7 ✠	4

Magn.	Stellarum characteres.	Ascens. recta	Num. Obseru.	Declinatio.	Num. Obseru.
7.8	.	177. 24. 10, :::	1	0. 24. 29, 4 —	1
7	.	177. 49. 20, 5	1	6. 55. 22, 6 ✠	1
5	♍ o	178. 11. 42, 6	2	10. 5. 20, 2 ✠	3
7.8	.	178. 30. 22, 1	1	1. 46. 13, 0 —	1
.	.	178. 52. 55, 6	1	1. 59. 0, 4 ✠	1
6	♍ r	179. 18. 6, 0	1	0. 27. 22, 0 ✠	1
6	♍ s	179. 24. 16, 5	2	7. 9. 50, 3 ✠	2
7.8	.	180. 26. 45, 8:::	1	4. 21. 58, 0 —	1
8	.	181. 32. 45, 1	1	0. 34. 17, 3 ✠	1
4.3	♍ η	181. 51. 37, 7	2	0. 41. 27, 6 ✠	3
5	♍ c	181. 59. 32, 7	2	4. 40. 26, 2 ✠	2
6.7	.	182. 32. 0, 7	1	6. 39. 53, 4 ✠	1
7	.	182. 40. 56, 5	1	3. 33. 27, 0 —	1
8.9	.	182. 57. 3, 1	1	9. 7. 31, 4 —	1
7.8	.	183. 50. 28, 4	1	3. 15. 39, 5 —	1
.	.	183. 58. 8, 9	1	5. 44. 58, 4 ✠	1
7	.	184. 23. 0, 8	1	12. 2. 12, 6 —	1
.	.	184. 35. 7, 2	1	2. 42. 34, 1 —	1
7	.	184. 46. 51, 9	2	3. 42. 11, 4 —	2
6	♍ q	185. 18. 14, 0	2	8. 6. 7, 7 —	3
.	.	185. 28. 35, 2	1	0. 3. 32, 3 —	1
6	♍ f	186. 3. 43, 3	2	4. 28. 59, 9 —	3
5	♍ χ	186. 40. 13, 4	2	6. 38. 53, 9 —	3
.	.	186. 44. 6, 1	1	4. 45. 13, 8 —	1
3	♍ 1 γ	187. 19. 44, 9	2	0. 6. 25, 3 —	2
.	♍ 2 γ	187. 19. 48, 7	2	0. 6. 30, 6 —	2
.	.	188. 44. 33, 0	1	4. 57. 46, 5 —	1
.	.	189. 23. 2, 1	1	6. 18. . . . . —	1
7	.	189. 40. 44, 3	1	9. 0. 13, 1 —	1
7	.	190. 10. 48, 1	1	2. 13. 16, 1 —	1
5	♍ ψ	190. 25. 26, 7	1	8. 12. 28, 5 —	1
3	♍ δ	190. 49. 56, 5	1	4. 43. 40, 4 ✠	1
6	♍ k	191. 46. 50, 0	2	2. 29. 23, 7 —	1
3	♍ ε	192. 30. 28, 8	2	12. 16. 35, 2 ✠	2
7	.	192. 50. 24, 8	1	2. 20. 33, 5 —	1
5	♍ g	193. 47. 15, 1	2	9. 25. 44, 5 —	2
4	♍ θ	194. 20. 16, 5	1	4. 13. 46, 3 —	1
.	.	194. 46. 45, 8	2	14. 52. 24, 9 —	2
7	.	195. 23. 15, 3	1	2. 45. 34, 0 ✠	1
.	.	169. 25. 18, 1	4	16. 56. 48, 5 —	6
4	.	196. 25. 43, 6	1	21. 52. 33, 2 —	1
9	.	197. 19. 9, 7	1	11. 17. 33, 1 —	1
.	.	197. 29. 55, 4	4	16. 26. 53, 7 —	6
1	Spica ♍ α	198. 5. 38, 1	...	9. 52. 45, 2 —	10
6	♍ i	198. 28. 1, 1	4	11. 25. 45, 5 —	4

Magn.	Stellarum characteres	Afcenf. recta.	Num. Obferu.	Declinatio.	Num. Obferu.
·	.	198. 37. 10, 3	4	14. 41. 55, 0—	4
7	..	199. 11. 21, 8	1	0. 5. 21, 1—	1
7	.	199. 26. 1, 6	3	5. 12. 4, 7—	3
5	♍ 1 l	199. 49. 43, 3	4	4. 59. 11, 8—	4
6	♍ h	200. 2. 19, 7	4	8. 53. 52, 7—	4
3	♍ ζ	200. 34. 22, 8	5	0. 39. 38, 5 ✠	5
6	♍ 2 l	200. 42. 56, 6	4	4. 8. 35, 1—	4
6	♍ m	202. 12. 38, 8	4	7. 27. 40, 8—	4
7.8	.	202. 48. 35, 2	1	4. 15. 32, 9—	1
7	.	202. 55. 35, 7	1	12. 58. 36, 5—	1
·	.	203. 14. 41, 6	4	11. 11. 35, 5—	4
·	.	203. 32. 59, 9	2	16. 37. 34, 2—	2
·	.	204. 9. 56, 8	2	16. 54. 24, 7—	2
2	Urſae maioris η	204. 28. 34, 9	...	50. 32. 20, 7 ✠	14
6	♍ p	205. 33. 8, 3	6	0. 17. 29, 2—	6
7	.	206. 48. 50, 6	1	6. 57. 44, 9—	1
7	.	207. 52. 8, 7	1	8. 4. 26, 2—	1
2	Centauri θ	208. 6. 18, 6	3	35. 9. 27, 4—	3
7.8	.	208. 21. 18, 0	1	7. 42. 53, 1—	1
6.7	.	208. 27. 41, 8	2	8. 8. 13, 2—	2
8	.	208. 29. 48, 8	1	15. 0. 35, 1—	2
6.7	.	209. 23. 26, 4	6	15. 8. 9, 6—	6
4	♍ κ	209. 58. 50, 2	6	9. 7. 29, 4—	6
·	.	210. 25. 18, 8	4	4. 48. 1, 0—	3
4	♍ ι	210. 48. 55, 0	5	4. 49. 24, 1—	6
1	Arcturus	211. 8. 1, 3	...	20. 27. 47, 2 ✠	29
4	♍ λ	211. 29. 16, 1	5	12. 14. 0, 8—	6
8	.	211. 46. 33, 5	2	5. 36. 29, 7—	2
8	.	212. 26. 38, 3	1	6. 38. 0, 8—	1
8	.	212. 35. 10, 6	4	10. 35. 6, 5—	4
8	.	212. 54. 12, 0	4	10. 32. 51, 8—	4
8	.	213. 1. 50, 6	4	12. 14. 4, 2—	4
8.9	.	213. 38. 32, 6	4	12. 14. 52, 8—	4
8	.	213. 53. 3, 3	4	8. 53. 40, 9—	4
7.8	.	215. 17. 28, 0	4	19. 21. 9, 9—	4
7.8	.	216. 0. 49, 4	5	11. 15. 11, 0—	5
7.8	.	216. 29. 0, 0	4	9. 29. 5, 9—	5
8	.	217. 12. 40, 0	4	11. 10. 31, 8—	6
4	♍ μ	217. 33. 40, 5	5	4. 34. 53, 9—	6
7.8	.	218. 5. 57, 7	1	20. 7. 23, 3—	1
7.8	.	218. 21. 12, 5	1	20. 16. 57, 0—	1
5	♎ μ	218. 59. 50, 5	5	13. 6. 54, 5—	6
2.3	♎ 1 α	219. 18. 34, 3	4	14. 57. 54, 5—	4
..	♎ 2 α	219. 21. 26, 4	6	15. 0. 38, 5—	6
6	♎ 1 ξ	220. 17. 39, 4	5	10. 53. 8, 0—	6

Magn.	Stellarum characteres.	Afcenf. recta	Num. Obferu.	Declinatio.	Num. Obferu.
6	♎ 2 ξ	220. 53. 35, 3	5	10. 24. 27, 3—	6
·	.	221. 15. 40, 7	3	10. 9. 28, 4—	3
6.7	.	221. 26. 6, 6	4	10. 8. 44, 1—	4
4.5	♎ δ	221. 59. 41, 3	5	7. 32. 0, 6—	5
3	♏ γ	222. 27. 44, 2	3	24. 18. 12, 0—	3
5	♎ ν	223. 15. 58, 3	3	15. 17. 30, 2—	3
7.8	.	223. 18. 41, 2	2	15. 31. 14, 8—	2
4	♎ ι	224. 35. 28, 3	3	18. 50. 54, 3—	3
7.8	.	224. 52. 3, 5	2	18. 42. 34, 8—	2
7	.	225. 34. 25, 5	1	21. 28. 33, 1—	1
2	♎ β	225. 58. 48, 2	4	8. 27. 51, 6—	5
9	.	226. 33. 46, 8	1	8. 14. 10, 6—	1
8	.	226. 46. 36, 5	1	17. 15. 10, 2—	1
7	♎ o	226. 51. 24, 8	4	14. 38. 56, 4—	4
6.7	.	227. 36. 15, 0::	1	11. 28. 45, 2—	2
4	♎ ε	227. 45. 12, 5	4	9. 25. 33, 4—	4
4	♎ 1 ζ	228. 38. 7, 3	4	15. 50. 44, 2—	4
·	.	228. 57. 19, 9	1	16. 34. 37, 6—	1
·	♎ 2 ζ	229. 14. 18, 9	4	15. 45. 6, 0—	4
7.8	.	229. 25. 54, 4:	1	19. 18. 36, 1—	1
7	.	229. 39. 41, 2	1	18. 49. 5, 4—	1
·	♎ 3 ζ	229. 47. 43, 0	4	16. 0. 18, 2—	4
·	.	229. 58. 23, 4	3	27. 12. 8, 0—	3
3.4	♎ γ	230. 28. 47, 3	3	13. 57. 19, 0—	3
·	.	230. 34. 6, 7	4	27. 18. 14, 5—	4
8	.	230. 43. 22, 4	1	25. 26. 56, 1—	1
9	.	231. 6. 14, 8:	1	13. 40 . . . —	1
·	.	231. 13. 37, 1	1	18. 28. 34, 5—	1
8	.	231. 13. 52, 3	1	13. 41. 26, 4—	1
7	.	231. 28. 40, 2	2	23. 0. 9, 2—	2
4	♎ κ	231. 59. 5, 1	4	18. 51. 58, 8—	4
8	.	232. 16. 21, 9	1	15. 12. 33, 4—	1
4	♎ η	232. 35. 51, 6	4	14. 52. 26, 7—	4
·	Serpentis α	233. 4. 9, 4	4	7. 12. 40, 7—	4
6	♏ b	234. 5. 20, 1	4	24. 59. 10, 5—	4
5	♏ A	234. 45. 10, 8	4	24. 34. 29, 1—	4
4	♎ λ	234. 48. 12, 1	4	19. 24. 53, 1—	4
4	♎ θ	234. 59. 35, 5	4	15. 59. 27, 0—	4
7.8	.	235. 0. 48, 7	1	24. 29. 47, 6—	1
8	.	235. 21. 51, 1::	1	18. 38. 22, 2—	1
4.3	♏ ρ	235. 28. 5, 2	3	28. 28. 37, 4—	5
3	♏ π	236. 2. 3, 6	5	25. 23. 14, 8—	5
·	.	236. 8. 28, 7	3	13. 33. 13, 1—	4
3.2	♏ δ	236. 29. 16, 1	5	21. 54. 12, 9—	5
7	.	237. 9. 42, 0	2	25. 9. 38, 4—	3

Magn.	Stellarum characteres.	Ascens. recta	Num. Obseru.	Declinatio.	Num. Obseru.
4.5	♏ ξ	237. 44. 49, 5	5	10. 40. 42, 0—	5
4	♏ 1 β	237. 49. 20, 5	5	19. 6. 52, 1—	5
8	♏ 2 β	237. 49. 29, 4	5	19. 6. 39, 6—	5
5	♏ 1 ω	238. 8. 40, 6	2	19. 59. 4, 0—	2
5	♏ 2 ω	238. 7. 7, 9	3	20. 11. 8, 2—	3
6	♏ 1 c	239. 18. 32, 3	4	27. 45. 24, 9—	5
6.5	♏ 2 c	239. 19. 48, 8	5	27. 16. 3, 3—	4
8	. .	239. 27. 32, 6	4	18. 47. 32, 1—	4
4	♏ ν	239. 27. 51, 0	4	18. 48. 11, 9—	5
8	.	240. 2. 8, 4	2	24. 50. 0, 8—	2
6.5	♏ 3 c	240. 48. 11, 9	5	27. 58. 51, 2—	5
.	.	241. 30. 1, 9	2	23. 33. 24, 1—	3
4	♏ σ	241. 36. 1, 7	6	24. 58. 55, 1—	6
8	.	242. 20. 43, 0	2	29. 6. 23, 5—	1
5	♏ ψ	242. 27. 58, 5	7	19. 26. 30, 9—	7
.	.	242. 44. 58, 4	6	22. 51. 35, 7—	6
5	♏ χ	243. 13. 48, 3	7	17. 52. 44, 6—	6
1	Antares, ♏ α	243. 37. 19, 5	9	25. 51. 54, 8—	9
7.8	.	244. 6. 42, 2	1	25. 58. 57, 4—	1
5	♏ φ	244. 18. 9, 2	6	16. 3. 23, 7—	5
5	♏ ω	244. 25. 40, 9	6	20. 55. 8, 6—	5
4	♏ τ	245. 11. 0, 8	6	27. 40. 58, 3—	6
8	.	246. 5. 23, 9	1	17. 40. 8, 9—	1
8	.	246. 33. 24, 3	1	19. 54. 26, 1—	1
7.8	.	246. 47. 50, 8	1	17. 33. 38, 1—	1
6	.	246. 52. 26, 7	3	17. 14. 42, 9—	3
7	.	246. 53. 51, 4	1	19. 25. 58, 4—	1
7	.	247. 26. 45, 8	3	28. 1. 46, 9—	3
.	.	248. 36. 5, 1	2	33. 49. 29, 8—	2
8	.	248. 43. 44, 4	2	24. 11. 6, 8—	2
8	.	249. 13. 48, 7	1	16. 6. 6, 7—	1
7	.	249. 45. 41, 7	2	19. 58. 55, 7—	2
.	.	250. 20. 1, 0	2	32. 50. 41, 2—	2
6.7	.	250. 31. 47, 5	1	22. 44. 10, 9—	1
7.8	.	250. 41. 22, 6	1	19. 7. 45, 0—	1
7	.	251. 15. 51, 1	2	24. 41. 40, 8—	2
6.7	.	251. 18. 40, 5	2	24. 35. 41, 9—	2
7.8	.	251. 54. 11, 8	1	18. 30. 7, 3—	1
8	.	252. 15. 19, 9	2	25. 19. 25, 3—	2
8	.	252. 26. 8, 0	2	25. 16. 14, 0—	2
7.8	.	252. 35. 3, 0	1	20. 7. 37, 1—	1
7.8	.	252. 54. 53, 5	1	21. 12. 0, 0—	1
7.8	.	252. 56. 55, 3	3	26. 9. 20, 0—	3
6.7	.	253. 31. 31, 8	3	17. 15. 37, 7—	3
3	Ophiuchi η	254. 6. 12, 4	3	15. 23. 50, 8—	3

Magn.	Stellarum characteres.	Ascens. recta	Num. Obseru.	Declinatio.	Num. Obseru.
5	Ophiuchi A	255. 5. 36, 0	4	19. 11. 29, 1—	4
.	.	255. 18. 49, 1	1	26. 10. 0, 8—	1
.	duplex .	255. 47. 19, 4	3	23. 59. 37, 6—	3
8	.	255. 48. 37, 2	3	23. 46. 33, 5—	3
4	Ophiuchi ρ	256. 36. 4, 2	2	20. 49. 30, 9—	2
8	.	256. 41. 17, 0	1	24. 41. 25, 4—	1
3	Ophiuchi. θ	256. 45. 44, 5	4	24. 43. 42, 2—	4
7	.	257. 0. 24, 9	1	27. 52. 46, 5—	1
.	.	257. 32. 2, 2	1	21. 11. 13, 2—	1
7	.	257. (33/48.) 10, 1	2	23. 59. 36, 1—	2
5	Ophiuchi b	257. 52. 22, 2	3	23. 55. 26, 0—	3
5	Ophiuchi c	259. 8. 12, 8	2	23. 44. 46, 9—	3
2.3	♏ λ	259. 16. 5, 4	1	36. 53. 50, 2—	1
.	.	259. 42. 6, 9	1	17. 17. 37, 4—	1
8	.	260. 9. 45, 8	1	21. 51. 7, 8—	1
7	.	260. 54. 20, 5	2	15. 23. 46, 4—	2
2	Ophiuchi α	260. 54. 26, 8	3	12. 45. 27, 4 ✠	4
.	.	260. 54. 31, 6	2	15. 13. 9, 9—	2
3	Draconis β	261. 14. 38, 2	1	52. 29. 31, 3 ✠	.
7	.	262. 10. 13, 0	1	15. 24. 44, 8—	1
6	.	262. 12. 42, 0::	1	21. 32. 13, 0—	1
6	♐ p	263. 3. 15, 3	3	27. 42. 36, 0—	4
.	.	263. 18. 13, 1	1	26. 51. 28, 3—	1
7.8	.	264. 5. 1, 9	1	22. 49. 3, 6—	1
7.8	.	264. 29. 16, 1	1	19. 1. 49, 7—	1
6	.	264. 52. 45, 0	1	10. 48. 48, 4—	1
6	.	265. 8. 14, 3	1	11. 15. 36, 1—	1
6.7	.	265. 23. 42, 0	2	18. 43. 59, 0—	2
7	.	265. 25. 16, 1	1	20. 52. 57, 2—	1
6.7	.	265. 34. 24, 9	2	15. 44. 32, 5—	2
6	♐ b	266. 13. 36, 8	3	23. 45. 58, 0—	4
.	.	266. 18. 36, 6	1	24. 14. 6, 9—	1
7.8	.	266. 22. 25, 8	1	20. 17. 35, 6—	1
6.7	.	266. 46. 58, 5	2	22. 44. 47, 1—	2
.	♐ 1 κ	266. 58. 39, 6	2	24. 15. 10, 9—	2
8	.	267. 3. 53, 2	2	22. 41. 25, 1—	2
.	♐ 2 κ	267. 13. 49, 2	2	24. 20. 15, 2—	2
3	♐ γ	267. 32. 7, 7	1	30. 23. 56, 3—	2
2	Draconis γ	267. 44. 35, 0	...	51. 31. 42, 2 ✠	.
7.8	.	268. 8. 30, [illegible]	1	21. 26. 30, 8—	1
6	.	268. 9. 31, 4	1	28. 27. 16, 3—	1
6.7	.	268. 35. 28, 2	2	30. 44. 17, 5—	2
6	.	269. 12. 29, 2	3	23. 43. 21, 6—	3
4	♐ 1 μ	269. 47. 43, 3	3	21. 5. 44, 5—	5
7	.	269. 54. 8, 5	2	21. 45. 0, 8—	2

Magn.	characteres	Ascens. recta.	Num. Obseru.	Declinatio.	Num. Obseru.
4	♐ 2μ	270. 9. 59, 8	5	20. 46. 25, 3—	6
7	.	270. 10. 30, 8	2	20. 25. 59, 4—	3
3	♐ δ	271. 20. 32, 0	5	29. 54. 10, 0—	7
2.3	♐ ε	271. 59. 37, 7	5	34. 28. 13, 6—	6
6	.	272. 42. 20, 4	2	20. 38. 42, 2—	2
4	♐ λ	273. 13. 38, 9	4	25. 31. 43, 8—	6
7.8	.	273. 34. 42, 9	3	25. 22. 59, 8—	3
7	.	273. 57. 51, 9	4	18. 51. 27, 4—	4
6	.	274. 15. 8, 2	3	19. 2. 9, 6—	3
7	.	274. 17. 9, 0	4	18. 32. 35, 5—	4
8	.	274. 32. 1, 5	3	19. 7. 14, 6—	3
6.7	.	274. 41. 51, 2	1	15. 1. 4, 4—	1
.	.	274. 42. 48, 1	1	18. 31. 8, 0—	1
7	.	274. 44. 39, 7	1	24. 11. 19, 4—	1
8	.	274. 54. 12, 1	1	24. 22. 58, 0—	1
6.7	.	274. 55. 40, 7	1	15. 0. 38, 0—	1
7	.	275. 15. 3, 3	3	19. 25. 58, 1—	3
7	.	275. 41. 10, 3	2	19. 23. 5, 9—	3
.	.	275. 49. 23, 1	1	21. 34. 21, 4—	3
7	.	275. 55. 2, 7	1	23. 41. 10, 0—	1
7.8	.	275. 55. 23, 7	3	17. 24. 42, 7—	3
7	.	276. 5. 5, 0	3	21. 12. 25, 8—	3
1	Lyra .	277. 10. 2, 5	...	38. 34. 16, 4 ✠	14
4	♐ φ	277. 36. 3, 0	4	27. 12. 51, 0—	6
7.8	.	277. 54. 9, 6	1	19. 50. 5, 7—	1
8	.	278. 20. 41, 5	2	20. 30. 40, 4—	2
.	.	278. 47. 48, 3	5	20. 34. 32, 2—	7
8	.	279. 2. 35, 7	2	22. 24. 56, 2—	2
8	.	279. 22. 14, 3	2	22. 10. 48, 1—	2
.	.	279. 51. 12, 5	1	21. 37. 58, 1—	1
5	♐ 1ν	279. 51. 30, 2	6	23. 1. 7, 1—	7
3	♐ σ	280. 1. 52, 3	5	26. 34. 19, 6—	6
5	♐ 2ν	280. 5. 26, 6	6	22. 56. 57, 3—	7
7	.	280. 42. 31, 9	3	20. 56. 55, 7—	3
4	♐ ξ	280. 47. 31, 9	5	21. 24. 1, 0—	6
8	.	280. 55. 37, 7:	1	20. 43. 21, 5—	1
6.7	.	281. 43. 45, 3	1	23. 0. 47, 2—	1
3	♐ ζ	281. 46. 2, 9	6	30. 12. 4, 0—	9
7	.	282. 9. 21, 0	1	25. 9. 21, 6—	1
4	♐ ο	282. 30. 48, 9	6	22. 4. 22, 3—	8
4	♐ τ	282. 55. 23, 1	6	27. 59. 59, 1—	7
7	.	283. 1. 57, 0	1	28. 58. 58, 3—	1
8	.	283. 4. 18, 3	1	22. 50. 45, 5—	1
7.8	.	283. 19. 50, 8	3	25. 0. 37, 0—	3
3.4	♐ π	283. 48. 40, 7	6	21. 23. 7, 8—	8

Magn.	characteres.	Ascens. recta	Num. Obseru.	Declinatio.	Num. Obseru.
7.8	.	283. 51. 4, 1	3	20. 9. 46, 6—	3
8	.	284. 28. 11, 4	1	22. 2. 13, 6—	1
8	.	284. 30. 24, 1	1	14. 57. 52, 4—	1
6.7	.	284. 32. 49, 8	2	26. 17. 13, 8—	2
6	♐ ψ	285. 8. 29, 0	3	25. 39. 2, 7—	6
7	.	285. 10. 6, 9	1	24. 34. 5, 8—	1
9	.	285. 25. 29, 8:	1	17. 44. 31, 8—	1
6	♐ d	285. 50. 16, 0	4	19. 21. 43, 2—	6
8.9	.	285. 58. 36, 8	2	19. 16. 32, 2—	2
6	.	286. 16. 13, 9	1	15. 56. 5, 6—	1
7	.	286. 30. 0, 9	2	22. 49. 49, 8—	2
5	♐ 1ρ	286. 52. 40, 5	3	18. 16. 52, 3—	5
6	♐ 2ρ	286. 54. 4, 0	3	18. 44. 9, 7—	6
6	♐ υ	286. 56. 11, 5	2	16. 23. 17, 4—	3
5	♐ 1χ	287. 36. 10, 0	3	24. 57. 23, 0—	3
5	♐ 2χ	287. 40. 40, 6	4	24. 24. 48, 5—	5
7	.	287. 56. 33, 1	1	22. 14. 3, 0—	2
7	.	288. 4. 18, 4	1	15. 34. 14, 8—	1
7.8	eadem forsan cum 287. 56.	288. 12. 29, 0	.	22. 14. 3, 0—	2
7.8	.	288. 40. 31, 8	2	15. 34. 27, 8—	2
.	.	288. 41. 17, 4	1	27. 27. 26, 7—	2
8	.	288. 54. 47, 6:	1	15. 50. 9, 7—	1
8	eadem videtur .	289. 6. 6, 0:	1	21. 47. 37, 9—	2
7.8	.	289. 12. 1, 0	1		
8	.	289. 57. 18, 3	4	24. 21. 39, 8—	4
5	♐ 1h	290. 17. 54, 3	4	25. 13. 40, 7—	5
5	♐ 2h	290. 27. 36, 9	4	25. 23. 50, 7—	6
8	.	290. 32. 44, 2	1	19. 21. 56, 8—	2
8	.	290. 43. 17, 0	1	18. 44. 57, 2—	1
8	.	291. 17. 4, 2	2	23. 57. 28, 6—	1
7.8	.	291. 21. 28, 8	2	23. 57. 39, 2—	1
6	♐ 1e	291. 40. 58, 2	10	16. 49. 43, 8—	13
6.5	♐ 2e	292. 8. 12, 3	9	16. 40. 20, 2—	13
6	♐ 3e	292. 24. 25, 1	9	16. 0. 43, 3—	12
6	♐ f	293. 1. 42, 6	9	20. 19. 25, 0—	11
8	.	293. 24. 2, 8	1	21. 31. 59, 6—	1
7.8	.	293. 34. 57, 6	1	14. 16. 50, 3—	1
.	.	293. 48. 47, 5	1	12. 54. 8, 4—	1
1	Aquilae lucida	294. 43. 10, 9	...	8. 14. 36, 6 ✠	29
5	♐ ω	295. 12. 57, 5	2	26. 55. 17, 3—	2
5	♐ b	295. 29. 9, 9	2	27. 47. 29, 5—	6
5	♐ A	296. 0. 54, 8	2	26. 49. 52, 6—	3
5	♐ g	296. 1. 24, 7	3	16. 6. 56, 5—	5
6	♐ c	296. 54. 19, 9	8	28. 21. 51, 4—	14
7	.	297. 4. 18, 1	1	14. 17. 31, 2—	1

Magn.	Stellarum characteres.	Ascens. recta	Num. Obseru.	Declinatio.	Num. Obseru.
8	·	297. 18. 49, 0	1	23. 15. 25, 0—	1
8.9	·	297. 25. 48, 8	1	16. 4. 23, 3—	1
·	·	297. 38. 36, 8:::	1	21. 58. 48, 0::—	1
7.8	·	297. 54. 0, 9	2	12. 16. 6, 1—	2
7	·	297. 57. 47, 1	3	13. 19. 58, 8—	3
8	:	298. 31. 25, 6	5	19. 29. 15, 4—	4
8	·	298. 40. 27, 1	1	15. 42. 29, 6—	1
7.8	·	298. 47. 34, 6	1	10. 44. 52, 0—	1
·	·	298. 49. 4, 0	1	21. 16. 31, 9—	1
9	·	299. 3. 57, 1	1	20. 4. 15, 0—	1
6	♑ 1 ξ	299. 36. 22, 8	10	13. 5. 44, 9—	9
6	♑ 2 ξ	299. 42. 16, 9	10	13. 18. 40, 5—	10
4	♑ 1 α	300. 42. 48, 3	11	13. 3. 45, 0—	11
·	♑ 2 α	301. 1. 36, 7	10	13. 14. 27, 8—	11
3	♑ 3 α	301. 7. 28, 5	12	13. 16. 49, 0—	12
6	♑ σ	301. 19. 22, 8	4	19. 51. 28, 2—	4
6.7	·	301. 45. 41, 1	8	15. 32. 0, 0—	9
6	♑ ν	301. 46. 40, 0	6	13. 30. 22, 3—	7
3	♑ β	301. 49. 13, 8	8	15. 31. 49, 3—	9
7.8	·	302. 41. 15, 1::	1	15. 1. 2, 5—	1
7.8	·	302. 53. 48, 4	1	14. 52. 42, 1—	1
8	·	302. 55. 16, 8	1	12. 28. 33, 5—	1
·	·	303. 0. 27, 5	1	20. 12. 4, 4—	1
6	♑ π	303. 19. 58, 2	6	18. 59. 25, 2—	6
6	♑ ρ	303. 43. 46, 2	8	18. 35. 58, 3—	8
8	·	303. 45. 46, 2	7	18. 39. 19, 7—	7
8	·	303. 45. 56, 2	1	18. 13. 18, 2—	1
8	♑ 1 o	303. 57. 52, 4	2	19. 32. 19, 4—	2
6	♑ 2 o	303. 58. 15, 4	4	19. 32. 11, 9—	5
8.9	·	304. 34. 41, 2	1	25. 44. 46, 3—	1
8	·	304. 40. 13, 9	1	17. 24. 52, 5—	1
6	·	304. 45. 42, 7	1	10. 36. 26, 2—	1
7	·	305. 8. 48, 2	2	14. 31. 21, 3—	2
7.8	·	305. 25. 36, 1	3	17. 20. 33, 8—	3
8	·	305. 34. 17, 9	1	21. 24. 27, 7—	1
8	·	305. 54. 40, 5	2	15. 58. 17, 8—	2
8	·	305. 59. 28, 6	2	17. 57. 3, 2—	2
6	♑ τ	306. 24. 4, 6	5	15. 47. 26, 6—	5
6	♑ υ	306. 32. 0, 1	7	18. 58. 38, 2—	8
7	·	306. 42. 0, 8	1	16. 58. 17, 7—	1
5	♑ ψ	307. 54. 8, 3	6	26. 7. 37, 0—	5
1	Cygni cauda	308. 16. 36, 7	...	44. 25. 11, 6+	14
5	♒ ε	308. 36. 44, 1	5	10. 22. 13, 6—	7
7	·	309. 7. 56, 2	1	18. 54. 57, 5—	1
6	♑ ω	309. 18. 13, 3	6	27. 48. 40, 5—	7

Magn.	Stellarum characteres	Ascens. recta	Num. Obseru.	Declinatio.	Num. Obseru.
8	·	309. 45. 23, 8	1	14. 6. 9, 6—	1
4.5	♒ μ	309. 52. 6, 6	7	9. 52. 49, 4—	8
7	·	309. 55. 51, 9	1	12. 28. 39, 0—	1
7	·	310. 14. 43, 6	5	18. 49. 47, 0—	6
7.8	·	310. 33. 45, 0	1	27. 12. 20, 2—	1
7	·	310. 59. 47, 5	2	16. 57. 7, 8—	3
7.8	·	311. 25. 37, 1	4	19. 57. 44, 9—	5
7	·	311. 46. 31, 2	4	18. 27. 49, 9—	5
7.8	·	311. 55. 5, 4	1	14. 27. 56, 6—	1
7.8	·	312. 9. 53, 2	1	12. 38. 2, 2—	1
5	♑ η	312. 37. 18, 9	5	20. 48. 2, 5—	6
5	♑ θ	313. 3. 3, 4	7	18. 11. 4, 5—	8
6	♑ A	313. 12. 14, 9	5	25. 57. 44, 0—	6
6	♑ 1 χ	313. 38. 16, 6	7	22. 9. 19, 9—	8
6	♑ 2 χ	313. 53. 29, 5	7	21. 31. 2, 8—	7
5	♒ ν	314. 4. 12, 8	8	12. 20. 32, 9—	9
7.8	·	314. 32. 49, 3	1	15. 27. 6, 8—	1
7.8	·	314. 59. 36, 4	3	23. 11. 58, 5—	2
6	♑ φ	315. 25. 40, 7	6	21. 38. 47, 5—	6
6	·	316. 3. 37, 0	5	18. 59. 24, 5—	5
7.8	·	316. 8. 32, 3	3	21. 20. 23, 8—	3
7.8	·	316. 24. 57, 7	1	17. 11. 16, 1—	1
8.9	·	316. 46. 52, 0	1	21. 49. 58, 4—	1
5	♑ ι	317. 9. 32, 3	7	17. 51. 22, 9—	7
8	·	317. 31. 56, 9	1	23. 46. 29, 1—	1
7	·	317. 34. 22, 9	6	21. 52. 17, 9—	6
6	·	317. 42. 30, 6	1	13. 54. 33, 7—	1
5	♑ ζ	318. 10. 32, 0	7	23. 26. 50, 8—	7
6	♑ b	318. 41. 38, 6	6	22. 51. 9, 8—	6
7	·	318. 44. 54, 3	1	12. 36. 36, 8—	1
7.8	·	319. 3. 58, 0	5	20. 11. 44, 9—	5
8	·	319. 19. 12, 8	2	15. 20. 36, 1—	2
9	·	319. 25. 24, 0	1	20. 17. 29, 4—	1
3	♒ β	319. 40. 32, 7	6	6. 37. 46, 6—	6
9	·	320. 2. 23, 9	1	17. 15. 21, 5—	1
6.7	·	320. 16. 42, 8	5	21. 9. 20, 5—	5
·	·	320. 17. 47, 2	1	21. 19. 0, 0—	1
4	♑ ε	320. 50. 52, 0	5	20. 32. 34, 8—	5
5	♒ ξ	321. 11. 14, 8	5	8. 56. 0, 6—	5
4.3	♑ γ	321. 38. 11, 0	5	17. 44. 56, 5—	5
·	·	322. 3. 50, 3	5	15. 7. 13, 8—	5
5	♑ κ	322. 15. 3, 1	4	19. 57. 47, 3—	4
7.8	·	322. 23. 11, 5	2	20. 43. 7, 3—	2
6	♑ c	322. 59. 42, 3	4	10. 11. 19, 0—	4
5	♑ λ	323. 20. 43, 7	4	12. 28. 36, 8—	4

Magn.	Stellarum characteres	Afcenf. recta.	Num. Obferu.	Declinatio.	Num. Obferu.
3	♑ δ	323. 23. 15, 4	3	17. 13. 10, 6 —	4
7.8	.	324. 7. 12, 3	2	13. 50. 46, 8 —	2
8	.	324. 11. 31, 6	2	17. 58. 3, 7 —	2
8	.	324. 31. 46, 4	2	19. 44. 42, 6 —	2
5	♑ μ	324. 59. 34, 4	3	14. 41. 10, 4 —	5
8	.	325. 25. 17, 1	1	16. 23. 43, 9 —	1
8.9	.	325. 49. 28, 9	2	19. 2. 32, 0 —	3
9	.	326. 7. 17, 2	1	16. 16. 16, 6:: —	1
7	.	326. 16. 19, 4	1	22. 20. 2, 7 —	1
8	.	326. 18. 25, 0	1	13. 49. 11, 4 —	1
8	.	327. 11. 8, 1	2	19. 3. 39, 0 —	3
5	♒ ο	327. 40. 21, 5	4	3. 19. 16, 5 —	3
4	♒ ι	328. 18. 33, 3	3	15. 2. 25, 7 —	3
3	♒ α	328. 18. 49, 2	3	1. 29. 36, 3 —	3
7.8	.	329. 22. 24, 3	1	12. 0. 30, 7 —	1
6	♒ e	329. 23. 26, 8	3	12. 45. 8, 4 —	4
7.8	.	330. 5. 35, 8	1	13. 7. 10, 7 —	1
4	♒ θ	330. 59. 12, 2	2	8. 59. 11, 5 —	2
7	.	330. 59. 39, 2	1	10. 14. 38, 8 —	1
5.6	♒ ρ	331. 50. 15, 2	2	9. 2. 6, 4 —	3
3	♒ γ	332. 15. 49, 5	2	2. 36. 24, 9 —	2
6	.	332. 49. 59, 6	1	6. 3. 36, 3 —	1
7	.	333. 0. 6, 1	1	2. 24. 54, 2 —	1
7.8	.	333. 25. 45, 4	1	12. 27. 26, 9 —	1
4	♒ ζ	334. 4. 6, 6	2	1. 15. 27, 4 —	2
5	♒ σ	334. 25. 45, 7	2	11. 54. 58, 5 —	3
7.8	.	334. 41. 1, 6	1	12. 8. 47, 4 —	1
4	♒ η	335. 42. 21, 1	2	1. 21. 55, 0 —	2
5	♒ κ	336. 16. 42, 1	2	5. 27. 36, 8 —	2
8	.	336. 48. 41, 4	1	10. 37. 23, 3 —	1
8	.	336. 52. 6, 9	1	7. 47. 41, 4 —	1
8	.	337. 32. 36, 4	1	11. 22. 17, 1 —	1
7	.	337. 33. 57, 2	1	9. 34. 43, 1 —	1
8.9	.	337. 36. 34, 8:::	1	8. 29. 3, 0 —	1
7	.	337. 37. 20, 3	1	8. 13. 53, 1 —	1
9	.	338. 7. 38, 7:::	1	10. 55. 12, 3:: —	1
6	♒ 1 τ	338. 40. 58, 8	3	15. 20. 59, 6 —	3
6	♒ 2 τ	339. 9. 47, 7	4	14. 52. 17, 7 —	4
.	.	339. 30. 40, 2	1	8. 35. 43, 8 —	1
4	♒ λ	339. 58. 9, 8	5	8. 52. 10, 8 —	5
3	♒ δ	340. 25. 12, 3	5	17. 6. 35, 9 —	4
1	Fomahand	341. 1. 46, 0	...	30. 54. 24, 9 —	6
7	.	341. 53. 39, 8	2	10. 10. 49, 2 —	2
6.7	.	342. 10. 32, 0	2	8. 20. 49, 4 —	2
4	♓ β	342. 52. 4, 1	5	2. 30. 46, 2 ✠	4

Magn.	Stellarum characteres.	Afcenf. recta	Num. Obferu.	Declinatio.	Num. Obferu.
2	Scheat, Pegasi β	342. 59. 37, 7	6	26. 45. 56, 4 ✠	4
.	♒ h	343. 6. 28, 8	2	9. 0. 12, 3 —	2
7	.	343. 9. 3, 0	1	9. 5. . . . —	1
2	Marcab, Pegasi α	343. 9. 21, 5	6	13. 53. 54, 8 ✠	5
6.7	.	343. 17. 34, 2	1	9. 14. 50, 8 —	1
9	.	343. 37. 33, 9:	1	9. 1. 6, 4:: —	1
6	♓ A	344. 2. 54, 8	7	0. 48. 20, 3 ✠	7
7	.	344. 13. 50, 6	1	0. 49. 36, 0 ✠	1
8	.	344. 30. 40, 0:::	1	7. 16. 43, 2 —	1
5	♒ φ	345. 25. 15, 0	7	7. 21. 29, 3 —	7
5	♒ 1 ψ	345. 46. 31, 5	7	10. 24. 38, 5 —	7
6	♒ χ	346. 2. 58, 7	7	9. 3. 4, 7 —	7
4	♓ γ	346. 8. 0, 5	4	1. 57. 18, 1 ✠	5
5	♒ 2 ψ	346. 18. 14, 2	7	10. 30. 30, 4 —	7
5	♒ 3 ψ	346. 33. 55, 4	7	10. 56. 16, 5 —	7
5.6	♓ b	346. 58. 53, 5	7	4. 3. 19, 0 ✠	7
6.7	.	347. 45. 22, 8	2	1. 2. 35, 6 —	3
7.8	.	347. 46. 38, 1	1	9. 47. 35, 9 —	1
.	.	348. 0. 20, 2	1	6. 2. 3, 0 —	1
5	♓ κ	348. 36. 34, 2	6	0. 4. 33, 1 —	6
6.7	.	348. 41. 25, 5	3	0. 12. 37, 1 —	3
5	♓ θ	348. 54. 8, 8	5	5. 2. 34, 1 ✠	5
7	.	349. 13. 50, 7	1	5. 51. 24, 8 —	1
8	.	349. 15. 3, 9	1	2. 22. 27, 6 —	1
7	.	349. 23. 58, 6	1	3. 7. 43, 3 —	1
7.8	.	349. 43. 52, 6	3	5. 24. 57, 4 —	3
7	.	349. 51. 43, 3	4	2. 25. 42, 9 —	4
6.7	.	350. 24. 13, 2	5	2. 35. 26, 3 —	5
7.8	.	350. 44. 13, 2	4	8. 48. 38, 0 —	4
7	.	350. 45. 16, 9	1	0. 1. 48, 0 —	1
6.7	.	350. 59. 15, 3	3	0. 45. 7, 6 ✠	3
5	♓ ι	351. 51. 17, 7	7	4. 18. 28, 8 ✠	6
5	♓ λ	352. 24. 10, 7	7	0. 26. 26, 3 ✠	6
8	.	353. 5. 46, 2	2	5. 50. 28, 7 ✠	2
6.7	.	353. 29. 9, 2	4	2. 8. 9, 9 ✠	4
6.7	.	353. 40. 28, 3	1	13. 15. 28, 9 —	1
6.7	.	353. 51. 10, 7	1	1. 0. 21, 2 ✠	1
7	.	354. 0. 4, 5	1	7. 42. 55, 7 —	1
7	.	354. 5. 15, 8	2	0. 51. 43, 6 ✠	2
6.7	.	354. 14. 43, 6	3	0. 16. 34, 0 —	3
6	.	354. 24. 45, 3	2	11. 19. 50, 0 —	1
7	.	355. 6. 3, 0	4	4. 30. 28, 6 —	3
6.7	.	355. 34. 34, 6	3	1. 14. 46, 5 —	2
6.7	.	355. 39. 55. 3	3	5. 43. 0, 5 ✠	2
6.7	1 □	356. 32. 52, 4	6	4. 54. 28, 7 —	4

Magn.	Stellarum characteres.	Ascens. recta.	Num. Observ.	Declinatio.	Num. Observ.
5	✳ ω	356. 42. 4, 2	6	5. 30. 49, 8 ☩	4
5	2 □	357. 19. 58, 4	6	4. 23. 1, 4 —	4
5.	3 □	357. 21. 46, 6	4	7. 22. 5, 1 —	4
•	•	357. 23. 8, 9	1	4. 7. 22, 0 —	1
5. 6	✳ c	357. 30. 42, 3	2	7. 7. 51, 9 ☩	2
5	4 □	358. 12. 47, 8	6	7. 4. 15, 5 —	4
2	Andromedae caput	358. 57. 20, 1	6	27. 44. 37, 6 ☩	4
8	•	359. 21. 33, 6	3	3. 55. 0, 5 —	1

Pro mutatione latit. fixar. saeculari.

Argum.

Longitudo stellae a o ♈

	0 ☩	1 ☩	2 ☩	
	6 —	7 —	8 —	
°	"	"	"	°
0	0, 0	22, 0	38, 1	30
1	0, 8	22, 7	38, 5	29
2	1, 5	23, 3	38, 9	28
3	2, 3	24, 0	39, 2	27
4	3, 0	24, 6	39, 6	26
5	3, 8	25, 2	39, 9	25
6	4, 6	25, 8	40, 2	24
7	5, 3	26, 5	40, 5	23
8	6, 1	27, 1	40, 8	22
9	6, 9	27, 7	41, 0	21
10	7, 6	28, 3	41, 3	20
11	8, 4	28, 9	41, 5	19
12	9, 1	29, 5	41, 8	18
13	9, 9	30, 0	42, 0	17
14	10, 6	30, 6	42, 3	16
15	11, 4	31, 1	42, 5	15
16	12, 1	31, 6	42, 7	14
17	12, 8	32, 2	42, 9	13
18	13, 6	32, 7	43, 0	12
19	14, 3	33, 2	43, 2	11
20	15, 0	33, 7	43, 3	10
21	15, 8	34, 2	43, 4	9
22	16, 5	34, 6	43, 5	8
23	17, 2	35, 1	43, 6	7
24	17, 9	35, 5	43, 7	6
25	18, 6	36, 0	43, 8	5
26	19, 3	36, 4	43, 8	4
27	20, 0	36, 9	43, 9	3
28	20, 7	37, 3	43, 9	2
29	21, 3	37, 7	44, 0	1
30	22, 0	38, 1	44, 0	0
	— 11	— 10	— 9	
	☩ 5	☩ 4	☩ 3	

☩ motum versus boream;
— motum versus austrum significat,
et signa haec valent pro tempore post epocham;
mutanda autem sunt pro tempore ante epocham.

VI. TREATISE ON THE PROPER MOTION OF THE STARS

When on a previous occasion a new catalogue of stars was described by me in this Society of yours,[244] gentlemen, I made mention at the same time of the proper motion exhibited by not a few of those stars which one usually calls 'fixed', as though they lacked all motion. And so I judged that when the opportunity arose for me to speak again, the features of this motion which I was able to observe would not be unworthy of your attention, the more so because they seemed to be capable of contributing to the perfection of this developing field of astronomy.

There is no need to go over all the discussions about this or similar motions of the stars which took place among earlier astronomers since, even though they may appear to have sensed some of this, I consider that they were deceived rather by the uncertainty of their own old observations, than by their reliance on the kind of argument that can be used to demonstrate or refute this motion. And perhaps it would not be difficult to show that all astronomers, who lived before this century of ours, admitted no movement in the stars which was peculiar to themselves, unless perhaps movement around the axis [of the diurnal rotation] but we are not speaking about that here; many did not even suspect [that there could be movement]. Although several phenomena were discovered by them which seem to relate to this matter, they are rather indications of those motions which we in our century have learnt are not really connected with the stars but are either aberrations of light or variations in the planes of the ecliptic and the equator, to which their positions are referred.

Unless I am mistaken, the first man who tried to demonstrate that some stars are not devoid of all movement which may cause them to increase or decrease their relative distances from one another was Halley,[245] who, after comparing the positions of Oculus Taurus, Arcturus and Sirius recently discovered by Flamsteed with those given in Ptolemy's catalogue,[246] did not hesitate to assert that these stars had advanced towards the south by about half a degree from Ptolemy's time to our own.[247] Then, in 1738, Cassini the Younger tried to deduce this same motion in Arcturus, and a similar one in the bright star of Aquila from more recent and therefore more reliable observations alone.[248] But no one worked with greater industry or provided clearer confirmation than the renowned le Monnier, when he detected some motion, albeit a very slow one, both in these and in certain other stars. For this purpose he used only more recent observations; namely his own, and those which Picard and [La] Hire made 50 years previously.[249] The method which he followed is described in the introduction to the Historia Coelestis published by him in 1741.[250]

But whatever these distinguished men have established so far concerning this matter must be regarded as the first probings in a new field, which it is our task and [that] of future astronomers to augment and perfect. For it is not enough to know that certain stars are endowed with their own proper motion, but we must also ascertain precisely which are the stars, in what region [of the sky] they move, and with what speed (constant or variable) since, unless we are certain about all these details, even the most accurate catalogue of stars cannot be employed with any degree of reliability over a long period of time, either for the past or for the future. But when these details have been established, no doubt many of the disagreements and obscurities which appear to have been thus far perceived in the observations of the ancient astronomers, will be easily overcome. But then other equally important advantages are also to be expected, for whose sake astronomers have sustained such great efforts in enquiring into the positions of the stars and in constructing their catalogues.

Many of these details must be left to posterity, since this motion of the stars is so slow that one age cannot suffice to define it accurately. However, since we have observations of our ancestors suitable for this task, it will be helpful to take these into consideration too and to use them to investigate this movement as far as is possible up to now.

You will not expect me, gentlemen, to set before you all those observations both of older and of more recent times which can be collated for the sake of learning about this movement, nor to subject all the stars to individual examination; this would be an enormous task. I shall only deal with such a section of the subject as it is feasible to discuss on this occasion; but from this it may be clear how large a harvest there is to be gathered in this field. Out of only 80 stars whose former positions I have been able so far to compare with their present-day ones, there are no fewer than 15 or 20 which, in the now sufficiently perceptible space of 50 years, were found to be endowed with their own proper motion. Amongst the remaining stars there are several for which one may suspect some movement, but it cannot be discerned on account of its slowness except over a longer time interval. Moreover, it is especially noteworthy that these wandering stars are to be found not only among those which belong to the first or second magnitude [classes] and whose movement may be seen to be faster since they are believed to be near to the Earth: they are also to be found among the less luminous stars, whose speed is no less than that of certain of the brighter ones, however much some of them seem to be completely at rest in space.

And indeed these and others of the same kind are manifest in the accompanying table [included in the original Latin edition], which contains two sets of right ascensions and declinations for more than 80 stars, reduced to a common epoch (i.e. 1756) except for those marked with an asterisk, which must be understood to refer to 1750. The right ascensions and declinations quoted in the columns were deduced with all possible care, even correcting for instrumental errors from the *Triduum* of that very accurate observer Roemer[251] which Horrebow obtained for me and inserted in his own works.[252] These observations were made in 1706, on 20, 22 and 23 October, exactly half a century before my similar observations of the fixed stars and the epoch of my new catalogue of them. Since this interval of time is not too narrow, and [since] scarcely any observations are to be found among other astronomers of that time which equal in accuracy those established by Roemer from his own eminently suitable instrument,[253] I therefore judged his [observations] to be especially suitable for learning how much the background of stars may have changed in the interim.

After those right ascensions and declinations deduced from the observations of Roemer there appear in the table similar positions of the same stars at the year 1756. These were copied from my catalogue and observed in the same and subsequent years, with the exception of the entries prefixed by an asterisk, which were taken from de la Caille's catalogue[254] since, on account of their being fainter and located outside the zodiac, the fixed stars to which these entries refer are not recorded in my catalogue.

And so this table refers, as it were, to two aspects of the sky: one, such as it was in 1706; the other, such as we now find it to be after 50 years.[255] In order that it may more easily be judged to what extent both [co-ordinates] agree or disagree, I have noted the difference in separate columns. This indeed, as is clear, is so slight in [the case of] most of the stars that, within the limits of observational error, it appears that they are really fixed and subject to no movement which can be perceived by our sense organs. However, in some cases the differences are a little greater than can be attributed to the imperfection of the instruments. Consequently, it would appear to be not improbable that those which exceed ten or 15 seconds of arc, and which on this account are distinguished from the rest, indicate a certain degree of movement. But in the case of Arcturus, Sirius, Procyone, Pollux, Lucida aquila, piscis austrinus and certain other stars, the difference between their former and present-day positions is so great, that there can be no doubt whatever concerning their movement.

In order that you may be able to discern the region into which any given star travels by [virtue of] its own movement, I have added the signs + and − for the differences. The former indicates a movement of the right ascension towards the east, [and] of declination towards the north; the latter, on the other hand, indicates a movement in right ascension towards the west [and] in declination towards the south. Thus, it is seen that Arcturus, the fastest of all, had advanced in the space of 50 years by almost two minutes of arc towards the south, and at the same time by one towards the west.

Finally, since the question may also be asked as to what is the cause of this movement, I am wont to utter this one word of warning; namely, that it cannot be explained by the movement of our whole solar system, although it is not impossible that the Sun moves through celestial space, since it is of the same nature as the stars and very like some of them. For if the Sun and all the planets together with it, including our own home the Earth, were moving straight towards some region, all the stars which appear in that region would seem to be gradually separating from each other one by one, and those which are in the opposite part of the sky would seem to be joining up; just as when you are walking through a wood the trees which are in front of you seem to be separated and those which are behind you seem to be joined together. Since these motions of stars are not bound by a common law of this kind, as is obvious from a closer inspection of the table, it is clear that they are not just illusory, or that they arise from this common cause or a similar one, but that the movements are peculiar to the stars. Perhaps the true and genuine reason for these movements will still remain unknown for many centuries.

COMMENTARY ON I

In basing his teaching on [the premise that] astronomical practice should be imitated in meteorology, Mayer was not acting heedlessly, but, as was his usual fashion, [only] after a careful consideration of the difficulties. He followed [that practice] not only in that method which he discussed, but also in his very procedure of carrying out observations and suitably recording them. It seems to me a great loss to natural philosophy that most meteorologists have not done this, for painstaking care is far from being employed [by them] in making observations, although the lofty purpose of the observations, i.e. [to acquire] a knowledge of the atmosphere and skill in predicting changes [in its condition], seems to demand it. Indeed, not even such care as we might be justified in demanding is usually devoted [to observations]. Anyone who takes the trouble of considering a little more closely the method by which these observations made up to now (I am referring, in particular, to thermometric observations) have been made by most of the men who have devoted themselves entirely to them, is unable to deny that this method is rarely if ever certain: it is concerned with any change in the state of the instrument, or in the atmosphere itself, and we must expect instrumental and observational errors. It often happens that observations of heat and cold taken in the same town and at the same time differ from each other to a degree that could scarcely be expected, other things being equal, from observations taken in remote regions.

This is of small wonder to anyone who knows that the majority of people who have devoted their efforts to these observations:

(i) have not even considered what is at the root of the matter, i.e. the determination of the limits of permissible error in either the instruments themselves or in their application;

(ii) have either not enquired at all or [have done but] in an unsatisfactory and inaccurate way, into the defects which can cause even instruments made by the most skilful craftsmen, to be unreliable;

(iii) have chosen places which are ill-suited for these observations, not taking that [degree of] care which the project demands;

(iv) either have not described the instruments they used at all, or [have done so] only perfunctorily; and

(v) did not describe the observations themselves with sufficient accuracy to ensure that one can have confidence in them, and form a correct assessment of their value.

In consequence, the result is that the thermometric observations are for the most part useless to those who want to investigate the general laws of heat and cold in a given country, and to draw conclusions from the given height of the thermometer about the temperature of the air surrounding that country, and to make comparisons with temperatures observed elsewhere. Before we decide upon the limits of the errors, which I know have sometimes been extended to the tenth degree of the Fahrenheit scale, and even further—and meteorologists are thinking of making them still smaller—[we should appreciate that] those conclusions drawn from individual observations are worthless since, when the sky (whose state we are discussing) is colder, the thermometric liquid can be observed at that time to be higher. I am not referring here to the mean annual temperature deduced from many observations, which indeed is nearer to the truth the more accurate these observations are but which can nevertheless not be much disturbed by erroneous observations, because the error is generally a random one, and so this mean temperature, according to the present state of the observations, will be the best measurement of the mildness and severity of the climate, and of all those phenomena which a comparison of the various types of observations will perhaps show to be functions, as it were, of this temperature.

There is no doubt that all the obstacles mentioned above can be removed and observations brought to a much greater degree of exactitude and self-consistency. But when will this come about? Undoubtedly when the meteorologists imitate the astronomers in their observations, since if the latter had followed the same method as the meteorological one, they would not yet have determined the altitude of the pole [i.e. the latitude] of any place, nor could they have ascertained the aberration of light and the proper motions of the stars. Indeed, the nature and application of meteorological instruments seem scarcely to admit such perfection as astronomical instruments now owe to English craftsmen in particular.[256] However, it is possible to envisage a greater degree of perfection than that with which the great majority [of meteorologists] have remained content hitherto, both from the experiments of the renowned Nollet[257] and Michael du Crest[258] and others, and especially from the outstanding work of the renowned de Luc, which he published in Geneva in 1772.[259] For in this last-mentioned work the great number of new experiments set up is described with the utmost exactness and indefatigable industry, and he [de Luc] applies a judicious modification to barometers and thermometers, so that no one, when he considers this work and the exhaustive labours of the distinguished men Toald,[260], Lambert,[261] and P. Cotte,[262] can doubt that the time for a more sensible [approach to] meteorology is now at last almost in sight.

à propos §6 and 10

The famous work which P. Cotte recently published in Paris has [in it] a table containing observations made at several places, of the maximum and minimum readings of both the barometer and the thermometer.[263] I have compared the mean heights of the thermometer calculated from this [table], with those given by a calculation made in accordance with §6 and 10 of [Mayer's] treatise. I have, incidentally, derived the altitudes of places above sea level from the mean heights of the barometer. The following table contains the differences [that] I found between calculation and observations. The thermometric readings are in Réaumur degrees.[264]

[TABLE 11]

Names of Places	Mean Temperature from observation	Mean Temperature from calculation	Difference
Mexico	13	12·7	−0·3
Chandernagore	20·8	19·5	−1·3
Cape of Good Hope	16·2	16·4	+0·2
Toulouse	11	11·6	+0·6
Florence	12·5	12·1	−0·4
Padua	11	11·5	+0·5
Paris	9	10·5	+1·5
Dunkirk	9	9·6	+0·6
The Hague	6·7	9	+2·3
Utrecht	8·7	9	+0·3
Edinburgh	6	7·6	+1·6
Stockholm	5	6·3	+1·3
Göttingen	8	9	+1

The mean height of the thermometer observed in Mexico City by D. Alzate was not taken from the table [on p. 386] but from §2, p. 340 of the same work, since the average from many readings seemed to me to be more accurate. That which is taken from the table gives a difference greater than −4 degrees.[265] It is to be regretted that we lack those observations which D. Alzate made in January, February and March, but their omission does not have such an effect on the determination of the mean temperature in tropical regions, as in our climate, because of the more stable state of the atmosphere. The mean heights of the thermometer observed at Göttingen and Paris, which I have listed in my table, also approximate more closely to those calculated [from §6 and 10 of Mayer's treatise], than to those which are (in fact) midway between the maximum and minimum [values]. I have taken the Paris reading from the treatise of P. Cotte[266] [and] the Göttingen [one] from the scale of Mayer's thermometer, described below.

There is certainly great agreement between the altitudes observed and those calculated; and on a stretch of land lying between the parallel of the Cape of Good Hope and Stockholm, and between the Stockholm and Mexican meridian, at an altitude of 858 toises above sea level (which according to the observation of D. Alzate[267] agrees with [that of] Mexico City), it seems to differ very little from the truth. But observations made outside these limits, e.g. at Petersburg and Vilna differ considerably from the calculations. The mean height of the thermometer which was observed at Petersburg and at Vilna was 8·5[268]—10·3 degrees less than that given by the formula.

Neither have we the time, nor is it appropriate here, to compare the [observed] mean altitudes of several places and indeed more accurate ones—there can be very little doubt about the mean values recorded above, since they were deduced from the maximum and minimum [temperatures]—with their calculated values, and with observations on other points in Mayer's theory; nor to evolve some kind of formulae for [deducing] corrections from them. These tasks must be left to others who intend to pursue this particular treatise of Mayer's further.

It remains for me to say something about the thermometer which Mayer constructed for himself in 1755 and always used afterwards in astronomical observations. This instrument is now in the hands of the distinguished Kästner, who procured it for himself after Mayer's death and very kindly got in touch with me so that I might add a brief description of it to this commentary. I do this with the greater pleasure because I know that it is of great significance to the literary world to have a more accurate knowledge of those instruments which distinguished men of letters have used in observations, and whose observations are regarded everywhere as the essential bases for calculation. For it quite often happens in this way, that what seemed to have been stated ambiguously in their writings can be established and explained, as I shall demonstrate presently with an example taken from this very thermometer of Mayer's.

The thermometer is a mercurial one, in a very narrow tube, with a bulb or oblong receptacle of mercury, about $\frac{2}{5}$ of a Paris inch wide and $1\frac{3}{4}$ [inch] long. The small wooden pole, to which the tube is fixed by the part nearest the bulb, has a little window, for the bulb itself, in the middle of which the latter is held very lightly longitudinally and latitudinally. The distance between the freezing and boiling points of water, or between +32 and +212 on the Fahrenheit scale, is around 13 inches. It has been graduated in six scales—those of Delisle, Fahrenheit, Réaumur the scale of the expansion of air, the Celsian scale, and finally the scale of mean annual temperature.[269] All are equal to each other in length (except the last-mentioned scale, which is shorter) and they are so arranged that the Réaumur and air thermometer scales are next to the tube, the Delisle and mean temperature scales are furthest to either side, and the Celsian and Fahrenheit scales are situated in the intermediate positions.

Concerning the Fahrenheit scale to which we shall first refer since it is more familiar to us than the rest which are now under discussion, we shall only note this fact: that it extends from −24 to +212, or from 24 degrees below freezing point up to the boiling point of water.* The scale of the air thermometer is divided into 500 equal parts. The freezing point is given the number 1000 and the boiling point 1380, but with the lowest point as I have just indicated, this is as low as −24 Fahrenheit (at the number 880). As far as I know, Mayer did not record the method for finding this; but there can be no doubt, since he was careful and painstaking in every field, that he checked these readings by experiments set up by himself. Moreover, the ratio 1000 : 1380 is no different from 1 : $\sqrt{2}$,[270] or from the ratio between the side of a square and a diagonal; which is equal to the ratio of the expansion of air, consistent with those degrees on the thermometer, as Bonne discovered.[271] But let us suppose, with Boerhaave[272] that air, when heated to the point of boiling water, expands to a third part of its volume—the air must of course be understood to be that which is halfway between hot and cold at Paris, because that rule was derived from the experiments of Amontons, which support it.[273] The following will then be quite obvious: at Paris and Göttingen, just as appears on the table set out above, the mean temperature is almost the same, i.e. +48 degrees Fahrenheit, but this corresponds exactly to the number 1034 on the scale of the air thermometer; if a third part of it ($=344\frac{2}{3}$) is added, the resulting total $=1378\frac{2}{3}$, which only differs by $\frac{4}{3}$ from 1380, the number which corresponds to the boiling point of water on the air thermometer.[274]

The scale of mean annual temperature extends from $+26\frac{1}{3}$ Fahrenheit to +70, its mean corresponding to +48. It is divided into seven parts pertaining to the months, and these are carefully prepared and disposed so that the means, as is also the case for the maxima, lie at equal intervals below and above +48; the rest, however, are smaller the further they are in either direction from the centre of the scale, the pairs [of values] which are at an equal distance from it being exactly equal. The middle portion [of the scale] covers temperatures for the middle of the months of April and October; the adjacent portion upwards refers to May and September, and downwards, to March and November; next to these, in the direction of increasing temperature, is the part for August, and in the direction of decreasing temperature, that for February and December. Of the two extremes, the maximum temperature has been fixed in the month of July, [and] the minimum in the middle of January. The following table

* *Translator's note:* Lichtenberg's familiarity with the Fahrenheit scale appears to have excluded the fact that −24°F is 56 degrees below freezing point.

shows in what order the months are arranged, and how many degrees Fahrenheit correspond to each one.

[TABLE 12]

+70	July = 5/6
	June/August = $5\frac{3}{4}$
	May/September = $9\frac{3}{4}$
+48	April/October = 11
	March/November = $9\frac{3}{4}$
	February/December = $5\frac{3}{4}$
$+26\frac{1}{3}$	January = 5/6

These seven separate parts are subdivided into further divisions, each comprising an interval of about five days, except for the two extremes which are triply divided because around the middle of July and January the temperatures remain stationary; since [the latter] follow the same law when they are decreasing as they obey when increasing, three subdivisions correspond exactly to six subdivisions of any other month. It is easy to discover the ratio and arrangement of these divisions and subdivisions from what has been said above and from the table set out in §15 of [Mayer's] treatise, by interpolation or by the construction of a mean temperature curve; as this can be done by those who wish to add similar suitable scales for the latitude of a place to their own thermometers, I shall make no further comment on the graduation of these. Likewise, I am omitting a description of Delisle's and Celsius's scales, but I must say a few words about [Mayer's] Réaumur thermometer, since its scale differs somewhat from that of conventional thermometers [of that type]. The temperature corresponding to +32 Fahrenheit is indeed 0 Réaumur, but to the degree +212 Fahrenheit it is neither +80 nor +100 Réaumur, but $82\frac{1}{2}$, although 0 Delisle and 100 Celsius coincide with the boiling point of water on the Fahrenheit scale. I shall not enquire into the reasons which led Mayer to choose this division: it is sufficient to record that the Réaumur degrees which he used in observations are rather smaller than the normal ones, of which 80 fall within the fixed Réaumur limits, and cannot be reduced to common Réaumur degrees, as is clear from a comparison of the tables in §7 of the treatise. For these 24 degrees Réaumur are taken to be approximately equal to 52 Fahrenheit, although 24 common Réaumur divisions are *exactly* equal to 54 Fahrenheit. I have mentioned the Réaumur scale of this thermometer both on account of the difference just stated, and especially because de Luc in the work to which I alluded favourably above included certain conjectures about Mayer's thermometer and its Réaumur scale.[275] I can venture to state here that these conjectures are considerably wide of the mark, the more confidently because the accuracy of this very wise man's remark is confirmed by this tiny discrepancy which occurs.

De Luc undoubtedly made a noteworthy discovery, worthy of the attention of physicists: namely, that there is a correspondence between the corrections to be applied according to Mayer's law for refractions for a change of 10 degrees on the Réaumur thermometer and those which ought to be applied according to his own careful observations for the same change to altitudes discovered with the help of the barometer and logarithmic differences. He afterwards made the following statement: 'This mutual confirmation', he said, 'of the experiments of M. Mayer and of myself, suggests that it is true that I have given a good explanation of the thermometer which this Astronomer used'.[276] It will not be irrelevant to demonstrate here with a short calculation how great this said agreement actually is. But in order that it may be both more easily expressed and understood, let us take the degree of the common Réaumur scale as $=r$, and the Réaumur degree according to Mayer's scale as $=m$, the degree of de Luc's thermometer (of which $23\frac{1}{4}=10r)=l$, the ratio $r : m=82{\cdot}5 : 80$ (just as it has on the scale)$=33 : 32$,[277] then $10r=93l/4=330m/32$, and so $m : l=93/4 : 330/32=124 : 55$ or $l=55m/124$. But according to Mayer, the variations of the thermometer$=10m$, the change of refraction$=1/22$ of the whole, therefore for a change l the change of refraction$=55/124{\cdot}220$ $=1/496$. But de Luc found that, with the change of the thermometer$=l$, the correction to be applied to altitudes deduced from the difference of the logarithm of the heights of the mercury barometer$=1/500$,[278] which is a fraction as much as 1/62 000 less than that which according to Mayer expresses the correction of the refractions for the same atmospheric temperature. De Luc, who put $m=r$, discovered a correction for refraction$=1/511$[279] which differs considerably from the 1/500 which has just been found; yet this difference is of no ultimate significance. One can only marvel that this remarkable agreement between the two is much greater than the subject demanded. At least it is not unpleasant to see such a small increase in that quantity.

COMMENTARY ON II

The learned author changed his original plan of adding a series of observations to this treatise, either [because] he was not able to set them out, being hampered by other affairs, or else [because] he changed [his mind] after the construction of the catalogue of fixed zodiacal stars. For of the pages which he had perhaps set aside for these observations, and had already marked out with lines for that purpose, only the first page contains certain zenith distances of the Sun and the fixed stars observed with the mural quadrant. I have omitted these here, both because the zodiacal stars have afterwards been listed in a catalogue, and also because the time of observation falls outside that epoch of the well-adjusted quadrant,[280] so that it is to some extent uncertain whether or not these are the observations to which the author is referring—especially when he does not agree with them elsewhere. Moreover, these observations will be published in due course, together with others selected by me from the author's diaries.

However, I have extracted the following [information] which the author promised on p. 15, from his diary. The zenith distances which are all

[TABLE 13]

With the quadrant turned towards the northern region.

Day of observation	Names of stars	Zenith distances					
22 July 1756	γ Drac[onis]	+0°	0′	0·2″	0°	0′	2·6″
	anonymous star 2	+0	10	15·6			
23	β Draconis	+1	0	7·6	0	57	55·2
	γ Draconis	+0	0	0·5½	0	0	7·2
24	β Draconis	+1	0	7·7½	0	57	57·1
	γ Draconis	+0	9	0·7	0	0	9·2
29	β Drac.	+1	0	7·7½			
	γ Drac.	+0	0	0·7	0	0	9·2
	anonym. 1	+0	7	1·7			
	anonym. 2	+0	10	15·9			
With the quadrant in its original position and nearly restored in [the plane of] the meridian.							
6 August 1756	anonym. 2	−0	10	15·5	0	38	33·7
7	β Drac.	−1	0	7·8	0	57	57·7
	γ Drac.	−0	0	7·9	0	0	9·2
8	β Drac.	−0	0	7·9	0	57	59·2
	γ Drac.	−0	0	0·6	0	0	7·8
	anonym. 1	−0	7	2·2: :	0	24	5·6: :
	anonym. 2	−0	10	15·6	0	38	35·0

apparent, and affected both by refraction and the apparent movement of the stars [i.e. parallax], are set out in the first column in 'degrees', one of which = 1/96 of the quadrant; the last one sets out their parts in the usual units mentioned on p. 14 above. What may be understood under the positive and negative zenith distances which the signs + and − signify, will easily be gathered at the very end of the observations.

In the diary itself there is another column after the zenith distances which contains the sum total of corrections for the heights of the barometer and thermometer. I have ignored this together with the actual numbers of those altitudes, because in all these observations it was = 0. Mayer used dots (e.g. this : :) to signify the degree of uncertainty in his observations; these dots are without doubt the numerators of fractions, of which the denominator = 8, for I did not see in the diaries a greater number of points than 8, which signifies the total uncertainty.

COMMENTARY ON IV

Although this treatise is now being brought to light for the first time, it no longer has the appearance of being original, since Lambert, having learned of this discovery of Mayer's in 1758 from new writings at Göttingen, was so struck by the originality and soundness of the idea that he explained and amplified it in such a way as to render it his own.* Referring the reader to that work, I would have readily abstained from all further exposition, had not both the triangles painted by Mayer, and the description of that which I have added to this work, demanded some kind of description from me.

But before I get down to this task, perhaps it will not be fruitless to consider in slightly more general terms a few facts from Mayer's theory, in order that the connecting links, together with the theory and purpose of these triangles, may be the more readily understood.

Therefore, let us imagine an equilateral triangle, divided into infinitely small areas—equilateral, because there is no obvious reason why we should take another shape. Each of the three areas at the vertices contains its own primary colour, r, b, g, the masses of which at the vertices themselves may be taken as $=1$. The strengths of colouring in the mixtures of these masses are proportional, so that, e.g. $m\mathrm{r}+m\mathrm{b}$† will yield a colour exactly half-way between r and b.

Moreover, throughout the remaining sections of the surface of the triangle, the colours r, b, g may be expressed in accordance with the same rule, so that the masses of those which merge in a given area are proportional to the perpendiculars ρ, β, γ from the area to the sides opposite the angles r, b, g. Thus, given the altitude of the triangle $=1$, the mass of red, which a given area will contain, will be $=\rho\mathrm{r}$, of blue $=\beta\mathrm{b}$, of yellow $=\gamma\mathrm{g}$, and the value of the mixture, or the colour of the area, can be conveniently expressed thus: $A=\rho\mathrm{r}+\beta\mathrm{b}+\gamma\mathrm{g}$, in which ρ, β, γ are positive variable quantities, whose sum (on account of the triangle [being] equilateral) is constant and always equal to the altitude of the triangle. Therefore, if one of these should disappear, the two remaining components [will produce] a lateral colour, or a mixture of two colours; because, if

* First of all in the Transactions of the Royal Academy of Science and Belles Lettres of Prussia for 1768,[281] then in a separate book, whose title is *Beschreibung einer mit dem Calauischen Wachse ausgemahlten Farben-Pyramide*, i.e. *Description of a Colour Pyramid painted with Calau wax* (i.e. discovered by the Berlin artist Calau). 4to [Berlin] 1772. See also Sulzer's Dictionary which is entitled: *Allgemeine Theorie der schönen Künste*. Art. FARBEN. i.e. General theory of the fine arts. Article on COLOURS.

† In these investigations, at least, I have preferred the method of expressing the proportion of colours in mixtures in which m is added as a factor to the signs of the colours.

one is taken as $=1$, or, which is the same thing, the sum of two $=0$, the equation will produce a primary colour.

Since there must exist three numbers, x, y, z, either rational or irrational, whose ratio can be expressed by three perpendiculars from a given point inside the triangle to [each of] the sides of the triangle, or by the ratio $\rho : \beta : \gamma$, it is clear that this triangle contains all the colours which can be mixed from these three at the vertices. No colour can be produced from a mixture, e.g. of colours mixed with compound or primary colours, which that triangle did not already contain. Indeed, no matter how many mixtures are blended together, the resulting mixture can always be reduced to the form $R\mathrm{r}+B\mathrm{b}+G\mathrm{g}$. Thus, when, for example, parallels are drawn both to the side opposite the angle r, and to the side opposite the angle b, the former at a distance $R/(R+B+G)$ from the former side, and the latter at a distance $B/(R+B+G)$ from the latter [side], the area in which these lines cut each other, will be that which contains the colour $R\mathrm{r}+B\mathrm{b}+G\mathrm{g}$. For the colour of that section will be $=R\mathrm{r}+B\mathrm{b}+G\mathrm{g}/(R+B+G)$, for which $\rho=R/(R+B+G)$; $\beta=B/(R+B+G)$; $\gamma=G/(R+B+G)$. But although this triangle may contain an infinite number of colours, it does not however contain absolutely every colour; the brightness and darkness of all the mixtures contained in it being contained within limits which cannot be transcended. Thus, for example, in our triangle, the colour $(\mathrm{r}+\mathrm{b}+\mathrm{g})/3$ which corresponds to the median section, will belong to the category of blacks, i.e. of those colours in which neither red, nor blue, nor yellow is predominant; but unless r, b, g are taken as being very dark, it will not be the colour which is commonly called black, but an ashen colour. The quality of blackness will vary according to the extent to which one, two, or all of these primary colours become darker.

What I have said about the ashen colour, holds true also for the rest of the colours; for all end in black itself, from which they can be recovered when white is mixed, just as ash can be produced from that black which we call pitch (which we see in burnt ivory) and many other coals.[282] These can be made lighter to some extent by placing coloured glasses between the eye and, e.g. a white wall, well illuminated by the sunlight. When the glasses are thick they block out light to the eye, and appear black, but when they are fine they appear tinged with colour; the finer the glasses are, the brighter [is this colour].[283]

Perhaps in our triangle there will be, for example, greens, some which you will call brighter, some darker. Since, however, they may be incorporated under the same formula $\beta\mathrm{b}+\gamma\mathrm{g}$, in which only β and γ are variables, they differ not only in brightness—just as, e.g. the colour of a glass plate of the thickness of $1/10$ of a line [$=1/120$ inch] differs from the colour of the same glass plate, whose thickness is one line [$=1/12$ inch]—but also in the ratio of the mixtures themselves. Thus, if you find the colour of one plate in our triangle, you will look in vain in it for the colour of the other; because in all plates of the same coloured glass, no matter what their thickness, the ratio $\beta : \gamma$ is the same. In order that all chromatic triangles may therefore be comprehended under the general equation set out above, which includes all possible colours, the colours r, b, g, must be taken as variables which can vary infinitely and grow darker, as it were, from extremely pale colours (equivalent to white) to the darkest colours (equivalent to black).

It will not be incongruous, since we consider the angular colours here as primary, to denote this variability with the help of explanatory symbols: thus r^n, b^n, g^n indicate colours nearer perfect darkness [i.e. black] than perfect brightness [i.e. white]; vice versa r^{-n}, b^{-n}, g^{-n} indicate colours nearer perfect brightness than darkness; while the symbols r^∞, b^∞, g^∞, and $r^{-\infty}$, $b^{-\infty}$, $g^{-\infty}$ indicate colours which are equivalent to black and to white respectively. Once this is established, it would be easy to write out a very general equation for infinite chromatic triangles, unless anyone should urge us to retain the equilateral triangle which we arbitrarily took above as a good method for the time being. One might take as a very general formula the following: $A = \rho r^n + \beta b^n + \gamma g^n$, for it will thus be clear that the whole system of colours can be set out as a prism whose bases are equilateral triangles. One, from the nature of the equation, [would be] of infinite brightness or whiteness, the other of infinite darkness, and the sections parallel to the base [would] satisfy all values (either negative or positive) of n itself.*

The triangle for which $n = 0$ in the general equation, is Mayer's chromatic triangle: it is identical to the section parallel to the base of the prism mentioned above which passes through the middle of the axis. This triangle contains all the colours exactly half way between white and black. Out of all the chromatic equilateral triangles, it displays the greatest variety of separate colours, and therefore in a general investigation such as this treatise contains it deserves particular attention. Lambert's triangle, whose angular colours are a very dark Berlin blue, a scarlet red, and the yellow of amber, and [which] can therefore be expressed in the form[285]

$$\rho r^m + \beta b^{m+p} + \gamma g^u$$

* The same result can be obtained more elegantly with two pyramids, exactly similar and equal and standing on [opposite sides of] the same base, provided that the numbers of the mixtures in the triangle which require to be distinguished, are taken as decreasing in the inverse ratio of the squares of the distances from the perfectly bright and [perfectly] dark ends, or in the ratio of n^2.[284] The similarity between this [representation] and what we have now placed on the triangle is clear from Mayer's treatise, and can readily be inferred from our equation. But these and various [other] propositions which can be derived even without calculation from a consideration of the profiles of these bodies, and which are not without use to the art of painting in general, I am leaving for another time and another place.

Plate 7

derives from a section exceedingly oblique to the axis of our prism. The field of both positive and negative exponent colours that can be individually weighted to an even greater extent by the mixture of white, has a very full, and hence for that reason, important use in the art of painting. In Mayer's system, however, when white is mixed, the painter can indeed produce all the colours of the negative exponents but none [associated] with positive exponents, since the black pigment (which indeed is not mixed *from* r, b, g, but mixed *with* r, b, g) does not raise those colours to higher powers. The black which can be mixed from r, b, g, and is comprehended by the expression $r^{\infty}+b^{\infty}+g^{\infty}$, when mixed with r, b, g, does not produce angular colours of a higher power, but merely produces dirty, squalid, faint, and hazy colours around the axis of the prism. Nevertheless, Mayer's triangle is eminently suited to the purpose for which its distinguished inventor had intended it; namely, to sketch out a general genealogy, as it were, of colours, and to enumerate all the colours which can be perceived. That this was indeed the object of the learned author, and was regarded by him as the key contribution is clear both from the treatise and also from the particular triangle which he constructed for himself, as we shall see later. For when he speaks about pigments, he seems only to have done so for the sake of illustration, in the manner of geometricians who illustrate what they are saying about lines with furrows drawn on a chalk table. But there is no doubt that the distinguished author, if he had lived, would have applied the general propositions of this treatise on colours, in practical terms, to the pigments for the use of craftsmen. Indeed, could anyone have done this better than Mayer, since this wise discoverer in the theory of art had also practised art itself from his boyhood with great success?[286]

According to the hypothesis which Mayer set out in his treatise (given that the altitude of the triangle or the sum $\rho+\beta+\gamma=12$, and the positive whole numbers ρ, β, γ), he tried to represent the 91 areas to which those infinites are reduced by this method, by pigments in a triangle which was handed to me along with the treatise. Its hexagonal sections, certain of which have been cut out of the appropriate paper and fixed in their proper places with wax in the manner of those which we have shown joined together in Tab. III (Plate 7), provide a model of a chromatic triangle, like some kind of patchwork. The angular pigments are cinnabar red, a paler mountain blue, and a yellow which we call royal, or artificial orpiment.[287] He himself in the new writings cited above,[288] when referring to his own discovery, defined the colours of those pigments as $=r^{12}$, b^{12} and g^{12}. He tried to reduce cinnabar and mountain blue to equal volumes with his own g, diluting them greatly with water, in order that the paper might transmit white. But although these are Mayer's angular pigments, he did not however obtain mixtures from both two and three colours with these pigments alone. In the sequence of greens we find that gay colour green bronze, which cannot by any means be derived from either a dry or a wet mixture of mountain blue and royal yellow. Similarly he did not mix his own pinks and violets from the angular pigments. Thus it seems to me probable that both for the primary colours, and also for the compound ones, Mayer sought the most suitable pigments; and where pigments themselves did not exist, he mixed the most suitable primary or compound colours. These movable areas made this task considerably easier, first of all, by frequently changing their positions, or by substituting some for others, whenever colour was passing into colour either too imperceptibly, or too rapidly or unevenly, it was immediately apparent which mixtures had still to be sought, and which had to be rejected from those already found. Then, when somewhat larger pieces of paper were tinged with some mixture, the areas could be cut out from those parts of them where the colour agreed and was uniform, for it happens not infrequently that one and the same pigment is produced from different parts of a painted piece of paper in a different way. For this reason, when you take from the paper a mixture which is good and suited to its position, it often does not fit into place any more. This is of great importance, if we are talking not about the triangle similar to ours, which we have given on Tab. III for the sake of illustration only, and an immense number of which must be completed for libraries; but about the unique one only, such as Mayer wished to display, which assuredly, as far as is possible, approaches the ideal of which we had a mental conception. That Mayer had great success in all these matters will not be doubted by anyone who is acquainted with the outstanding visual judgment in which he excelled, and who can see the remarkable evidence of this judgment in this triangle, or rather in these remains of the triangle (for one group of areas is lacking, and another has become faded, which upsets the harmony).

In this triangle the majority of the finest and most vivid colours are absent, since those of higher ranks are in other triangles in accordance with the theory established above. So, if Mayer had constructed his own 25 triangles, all the separate colours—as many, that is, as can be suitably represented by pigments (either pure or mixed)—would have been introduced into them. There would consequently have been as many chromatic triangles thus set out with pigments as there are pigments, and besides this affinity of colours, they could have constituted a kind of visual scientific catalogue. But I imagine that many areas, especially from the part [containing] the dark colours, would perhaps remain empty, since their colours cannot be mixed from known pigments, and the colours of many others cannot be discovered without the greatest difficulty. It is obvious that, as I have just indicated, we lack that dark pigment by whose help colours can be intensified[289] in the same way as we weakened those others to certain grades of white. This is because pigments which correspond, e.g. to the three angular colours, are very rarely constituted in such a way that their mixtures are obtainable from a [single] section of our prism, with the intersecting plane passing through the three given colours. They fall partly below, and partly above it, and even seem to be mixtures of the three primary colours, although they are [in fact mixtures] of two; also, they are subject to other changes—chemical ones in particular.

To §18–20

If you want to find—not by pure visual judgment and by experiment (for this cannot be done without a great loss of time), but rather by a direct mathematical method—all the colours of the triangle expressed in terms of their own pigments, and to convert [these] to the use of craftsmen, applying generally the broad principles which Mayer has supplied, you must investigate in particular the colouring strengths of the pigments and their mutual influence.

Since Mayer clearly did nothing about this, Lambert, in his own work praised above,[290] has consequently not only taught us a method of investigating these strengths, but also set out with considerable success to paint the colour pyramid. I shall select [only] certain important features of it, and indeed shall apply them immediately to the triangle which I have completed and described here, because this is not the place to give a detailed description of the experiments of the distinguished gentleman.

No doubt some friends who are skilled in these matters will ask me, even though it is impossible to add chequered and perfect triangles or *true chromatoscopes* for the reasons given above, to include some diagram in my commentary which would give a slightly better sensual representation of that unique triangle of Mayer's than the triangle where mere formulae with their individual distributives are written in the coloured areas [Tab. II]. I have tried this, therefore, in various ways, which it is superfluous to list here. I shall mention only two—one, in order to elucidate Lambert's method, the other because we have made use of it in painting the triangle on Tab. III.

At first I had the intention of painting the triangle in Tab. III with colours mixed from Mayer's three angular pigments, i.e. cinnabar, mountain blue, and royal yellow. I had decided to put these pigments suitably mixed into as many little containers as there are areas in the triangle. Since they were arranged in this way so as to form a replica of the triangle, the pigments could easily have been produced by anyone little skilled in the art of painting, and spread over their respective areas. But I was deterred from this plan not only by the dirty greens and violets which resulted from these mixtures, but in particular by the fact that, on account of the different specific gravities of the pigments, and the irregularity of their cohesion with water, large errors could easily be made even by someone skilled in art. Although mountain blue is often decanted and well stirred up with a brush before it is spread on the paper, it can nevertheless not be mixed with cinnabar or royal yellow in such a way that the mixture is consistent; and when drawn frequently from the same little container, it gradually changes from the colour which it possessed to begin with, and takes on another which does not correspond sufficiently accurately to the area to which it belonged. Moreover, although mountain blue must be given preference over the rest of the blue pigments because of its brightness and fineness, and especially because of its suitability, since it $=b^{12}$, yet it cannot be manipulated by anyone in such a way as to produce paper with a consistent colouring. And so, having abandoned my plan of painting a triangle of this kind for the bookseller, in order to satisfy my desire to see what mixtures can be produced from Mayer's angular pigments, I constructed a triangle from those dry pigments, as Mayer recommends at the end of §10, thus investigating their strengths of colouring by Lambert's method. With the help of a reliable set of scales, I obtained for myself small weights increasing in a double ratio, the least of which did not amount to more than the weight of a grain of wheat. When I had discovered portions of pigments (reduced to the finest dust) as accurately equal to these weights as I could make them, I mixed them with a thin dry brush. The following mixtures were obtained in this way. (The letters signify pigments, but the numbers indicate how many small weights any portion weighed; where a coefficient is not prefixed, they must be considered as having been multiplied by unity.)

$$g+b;\ g+2b;\ g+4b;\ g+5b;\ g+6b;\ g+7b$$
$$g+r;\ g+2r;\ g+3r;\ g+4r;\ g+5r$$
$$2r+b;\ r+b;\ r+2b;\ r+3b$$

I gave these mixtures to certain of my friends so that they might determine with their own eyes which of the mixtures fell exactly half way between the primary colours of which they were composed, or that they might indicate between which colours they thought the half way point fell. Combining the opinions [that they] voiced among themselves with my own, I came to the conclusion that the following facts could be established: that the mixture $g+6b$ gives a green in which neither yellow nor blue is predominant; similarly, that the mixture $g+2r$ or $g+3r$ is half way between g and r; and finally, that $r+2b$ is [half way] between r and b. There are, therefore, masses of these pigments which exercise an equal strength of colouring in mixtures:

$$g : b = 1 : 6$$
$$b : r = 2 : 1$$

from which is derived the ratio $g : r = 1 : 3$. Because this agreed with another visual judgment on these discoveries, I have retained it as the correct ratio. These masses of pigments $b : r : g$ are directly proportional to the numbers $6 : 3 : 1$, and thus the colouring strengths of equal masses are in a reciprocal ratio to them, or in the ratio $\frac{1}{6} : \frac{1}{3} : 1 = 1 : 2 : 6$. It is obvious that these ratios built on the values stated above apply only to those pigments which I used, i.e. [pigments] of the same nature and colour, reduced to a dust equally fine and equally dry, and mixed in the same way; for the slightest aberrations can produce greater differences than anyone who is not an expert can imagine. Moreover, it is utterly impossible to draw conclusions from these experiments regarding the ratios which would pertain to the strength of colouring of the same pigments combined with water or oil.

But once the ratio of the pigments' strengths of colouring, or the inverse ratio of those masses which produce some sort of equilibrium in mixtures, has been discovered and the ratio of the primary *colours* which make up any given mixture fixed, that colour can easily be mixed from the pigments, as we shall now show in a general formula. We shall suppose here that pigments differ from colours only in the inequality of their powers of colouring. E.g. let the three given pigments be R, B, G, whose colours are r^n, b^n, g^n. Their strengths of colouring are $1/m$, $1/n$, $1/p$, and the colour mixed from them is $\rho r^n + \beta b^n + \gamma g^n$. It is quite clear that the mixture must include a quantity of red pigment $m\rho$, of blue $n\beta$, of yellow $p\gamma$, and thus the required colour is produced by mixing the masses $m\rho R + n\beta B + p\gamma G$. In this formula, introducing Mayer's pigments and the numbers discovered above in place of m, n, p, and positing ρ, β, γ as constant positive integers, whose sum total always $=6$, the mixtures are produced which the following triangle contains. The letters r, b, g denote pigments, and are substituted for R, B, G for the sake of abbreviation; moreover, the numbers indicate the ratios of the portions of pigments which must be included in any given mixture in order to obtain a colour suitable to the position [on the paper]. I have omitted the distributives of the colours, because they are easily discovered by anyone who compares this triangle with that contained in Tab. II.[291]

[TABLE 14]

						R						
					$r^{15}b^{6}$		$r^{15}g^{1}$					
				$r^{12}b^{12}$		$r^{12}b^{6}g^{1}$		$r^{12}g^{2}$				
			$r^{9}b^{18}$		$r^{9}b^{12}g^{1}$		$r^{9}b^{6}g^{2}$		$r^{9}g^{3}$			
		$r^{6}b^{24}$		$r^{6}b^{18}g^{1}$		$r^{6}b^{12}g^{2}$		$r^{6}b^{6}g^{3}$		$r^{6}g^{4}$		
	$r^{3}b^{30}$		$r^{3}b^{24}g^{1}$		$r^{3}b^{18}g^{2}$		$r^{3}b^{12}g^{3}$		$r^{3}b^{6}g^{4}$		$r^{3}g^{5}$	
B		$b^{30}g^{1}$		$b^{24}g^{2}$		$b^{18}g^{3}$		$b^{12}g^{4}$		$b^{6}g^{5}$		G

I showed a triangle of this kind constructed from dry pigments to the Royal Society of Sciences [Göttingen] at a meeting which it held on 6 August this year.[292] Dry mixtures of mountain blue and royal yellow did not grow less faint than when diluted with water, and this was the case with most of the rest; for which reason, although these three pigments are resplendent with the finest colour in their primary state, they seem to be less suitable for mixtures. But if you wish to avoid major errors entirely, it is quite clear without my telling you that this method demands great accuracy. Once a mistake is made in the weighing, a pigment that contains considerable strength of colouring can cause such a disturbance in the mixture of the extremely tiny mass which is mixed, and hence in the ratios of the relative strengths of colouring deduced from it, as to render the whole operation almost worthless.

Now we come to the description of the method which we followed in painting our triangle. The angular pigments are natural cinnabar, a clearer Berlin blue and amber. We chose Berlin blue in preference to mountain blue, because it is easier to manage, and does not give rise to numerous spots which can scarcely be eradicated; we chose amber in preference to royal yellow, because it produces finer mixtures. They [i.e. these pigments] have the added advantage that their mixtures do not differ unduly from the mixtures of Mayer's triangle, of which we wanted to depict some sort of outline—provided that Berlin blue, with the white of the paper shining through, is reduced in some measure to the strength of mountain blue. These pigments were well decanted and watered down to the consistency of ink and seemed to have the same inherent transparency. When this was done, at first all the areas apart from the sides opposite the angle, were tinged with the red pigment; apart from the sides opposite the angle *b*, and apart from the sides opposite the angle *g*, all were tinged with yellow. In this way, the whole triangle appeared to be painted with seven colours: the angular sections, painted only once, contained the three primary colours; the areas lying between these sides, being painted twice, contained greens, violets and golds, all equidistant from their own two angles; the areas lying inside the surface of the triangle and between the sides, being painted three times, appeared as an ashen colour consisting of equal parts of the angular colours.*

* If, when this has been completed, the areas do not display the colours just mentioned, the latter must be brought to a kind of norm before proceeding any further either by adding some pigment, or by adding water to the mixtures mentioned (as we saw was done in Lambert's method of weighing) in order that as many masses of pigments proportional to their strengths of colouring might be put into those areas as are painted on top of them, ignoring the ratios of their strengths.

A second red was imposed, so to speak, upon the red triangle whose side was less than the side of the first by unity (i.e. by one area), but whose vertex coincided in *r* with the vertex of the first. Similar blues and yellows were then placed on the first blue and yellow triangles, until, with the triangles thus superimposed on one another continually growing smaller, and with their positions thus cyclically changing, all the areas comprising this pattern appeared coloured appropriately in accordance with the established theory; either on account of the transparency of the strata themselves, or the mixture of pigments made in the areas, or some other simultaneous cause. It is easily perceived that these coloured triangles were not superimposed upon the ones beneath them unless they were properly dry. With this task completed, a less experienced eye will see in what areas correction has to be made, since a comparison of all the areas made, as it were, at a single glance, shows the aberrations of the individual [pieces]. That sufficiently good triangles may be completed by this method can be tested by those who, having more than a passing acquaintance with the art of painting, want to prove this theory. Those [triangles] are not to be despised and, when the pigments have been properly diluted, are quite suited to the purpose at hand; for we see [that they] have been made even by boys with the minimal apparatus, where another method (especially if a triangle consisting of 91 areas had to be painted) would have demanded a great deal of apparatus. The great advantages of a triangle of this kind for artists, especially if it consists of only three suitable pigments, have been proved beyond all doubt in the work of Lambert.[293] Indeed, why should these advantages not be outstanding, when a single glance at all the colours which can be mixed from the three pigments enables one not only to consider their family relationship, as it were, but at the same time to visualize the method by which all the colours in the smallest mixture can be compounded from the primary colours? The amount of the portions may be found using scales—whether [this is estimated] visually or from a reliable pointer at the fulcrum of the scales, comes to the same thing. But artists do not readily permit themselves to be restricted by such narrow boundaries. All colours obtained from [mixing] three or four colours are produced in their finished state by four pigments; but since there are colours containing many pigments, which nature aided by art supplies to us, and which can scarcely be mixed from two or three pigments in such a way as to produce a colour equally vivid, equally rich, and equally lasting, the experiments set up concerning these triangles will be useful [only] to those artists whom the method of the art which they practise forces to work within these limits. I am referring in particular to the art of copying painted pictures in representations wrought out of bronze, [a technique] which I. C. le Blon (who died an old man in 1741 at Frankfurt, where he was born) discovered, and afterwards worked at in England and France.*

[The technique] is now known to only a few foreign craftsmen, which is a matter for great regret. The inventor tried to accomplish everything with three brass tablets. No doubt the individual ones were first of all roughened with a file, then with the help of a chisel and a smooth piece of steel, just as is wont to be done in the work which we call *Schwartzkunst*, and the French *manière noire*. They were scraped off again and polished to the shape of the figure to be represented and to the various degrees of light and shade. They were so arranged that, when any one was tinged with its own primary colour and stamped with the plate, it gave an imperfect copy of one colour; but when all were properly stamped with the same plate, and somehow placed on top of one another, with colours and shapes joined, they gave a perfect copy, painted with every colour. It is easy to see that this art is beset with a great many difficulties, and there is no need for me to demonstrate the fact. I cannot do better than select the passage in which the author of the French version of le Blon's book makes some remarks pertaining to this point, which will show at the same time the difficulty of the subject and the potentialities of the chromatic triangles made by le Blon's art: 'When one has come to make a model', he says on p. 107, 'one is well advanced; e.g. suppose that I have one portrait to engrave, and that there are a hundred different shades of colour. The coloured engraving of a St Peter, which I would have preserved, with the copper plates pressed upon it, determines one part of my shades, and this is how. I wish to colour the scarf of the portrait; this scarf appears to me, on inspection, to be of the same shade as the belt of my St Peter, which was painted long ago. I examine the copper engravings of St Peter, and realize that there is too much yellow and too much red in their texture. Then, to depict the scarf of the portrait, I reserve in yellow and in red as much texture as my ancient copper plates have in them for the belt of St Peter'.[294] Anyone who reads these words and compares them with what is discussed by the same author in other passages, will be amazed that he did not think of the method of displaying all the mixtures at the same time, and obtaining a sort of repertory, which the craftsman could consult whenever he was in doubt about a mixture. Moreover, who does not see that the chromatic triangle, or system of triangles, such as Lambert's

* He published in London a treatise written in English: *Il coloritto: or the harmony of colouring in painting reduced to mechanical practice and infallible rules* [1755], to which are said to be added certain specimens of his own art. I have not seen this edition, but the French version is to hand, to which (besides the English original) is added an exposition of the practical part of that art; but the specimens have been omitted. This [version] is entitled: *L'Art d'imprimer les Tableaux. Traité d'après les Écrits, les Opérations et les Instructions verbales de I. C. le Blon à Paris 1756. 8vo.* Many have written le Blond, but I thought that I should follow the authority of the French version.

pyramid affords, provides the repertory of this kind, i.e. the most complete, constructed in the least time and space, with the least effort and expense?*

For the craftsman constructing it for himself, the task will present only difficulties associated with others [already] overcome, and for that reason they will be more easily resolved; and when so many of these areas have been suitably represented for the purpose at hand, he will easily find the rest from them. Why? Surely it can thus be seen at a glance how much the craft and the craftsman are worth. If the latter is not equal to creating a perfect triangle from all the numbers, he will be much less able to paint pictures. It would be desirable that several craftsmen should employ their efforts in this craft to attempt to discover whether or not it would be possible to bring everything to a conclusion with three tablets (although four, and even five have been used by some) since there is almost no doubt that Lambert's pyramid could be constructed with three only—provided that it is permissible to apply conclusions from mixtures of painted pigments made on a palette with a painter's brush, to mixtures made from primary pigments with the thinnest layers superimposed upon each other and on paper.†

In painting it the distinguished gentleman [Lambert] used almost the same pigments that are usually employed in those copper-plate engravings. No one will deny that a table of a kind of negative white paper, i.e. one tinged with a copper-engraver's ink, was used not without the *occasional* loss of considerable elegance and grace.

Before I conclude these remarks of mine on this treatise, I cannot omit saying a few words about Mayer's discovery, which I thought would not come amiss to readers and would not be inappropriate at this point. For that learned gentleman, when thinking about his own chromatoscope or chromatometer, as he acknowledged in the new writings at Göttingen

* The nature of this pyramid can be conjectured from those which are mentioned at the end of §24 of Mayer's treatise. I only add this: that it consists of seven chromatic triangles arranged and delineated according to the rules of perspective in such a way that they seem to be different storeys of a little pyramidal box, not unlike those glass cases which are wont to rest in the corners of rooms; and the little areas contained in the individual [triangles] are quite easily seen. The base consists of a triangle of 45 areas, which we mentioned on p. 95; and rising towards the apex of the pyramid, are triangles of the following [number of] areas: 28; 15; 10; 6; 3; 1. Of these, the one nearest the apex contains perfect white.

† Anyone who wants to see the extent to which this can be done, should read the splendid works of the distinguished gentlemen *Gautier Dagoty*, father and son, especially those which they published most recently. I happened to see most of them, among which the following seemed to be particularly worth noting: *Cours complet d'Anatomie, peint et gravé en couleurs naturelles* [*Paris*] *1773*, and *Galerie universelle gravée en couleurs* [Paris] and that book which I have not yet seen, in which the distinguished author described his own method, entitled: *Système de l'Univers et de la Génération des Couleurs* [Paris]. An excellent specimen of this craft, unearthed by the inventor Le Blon, is preserved in our academy's library.

of the year 1759, p. 402, alighted upon the art of reproducing engravings of a given picture, clearly different from le Blon's method related above, and from Jackson's.[295] He showed the only example to the Royal Society of Sciences and moreover he concealed this skill as long as he was alive, and it could not be found after his death in his manuscripts. However, it must not be thought of as lost, since the son of the inventor [Johann Tobias], a man worthy of such a great father, and besides him the well-known Nuremberg artist [Georg Christoph] Günther, with whom Mayer was friendly, are still surviving, and we know that the chief principles, at least, of this art are known to them.*

In the specimen completed by the inventor himself, which his widow and son very kindly allowed me to examine a little more closely, I observed the following points. It seemed to be some sort of artistic work, in which coloured waxes were employed in place of glass or coloured marble; in such a way, however, that not only sticks or thin staffs and similar objects, but also whole wax plates, bent and curved to the various shapes of image to be created, are joined together with others of different shapes in a rectangular prism. The bases [of this prism] give the required picture, but the sections (however many there may be) are parallel to the bases, and yield other images similar and equal to them. From this prism the inventor cut off thin plates parallel to the base with a thin broad knife, and placed them on metal plates or on little beams. Such a specimen was also shown to me together with the prism of which it was a section. The height of the prism about which I am speaking here is now only $\frac{3}{4}$ inch, but I was told that it was originally three or more [inches]. The bases are rectangles about 6 inches broad and 4 [inches] high. They represent clear copies exemplifying *Guid. Reni*'s 'Erigone and Bacchus changed into a vine'. One can see, with the greatest delight, the wonderful, clear resemblance of the bases and section, since there is not the thinnest line or tiniest point which is not seen to correspond exactly in all respects, and this continues throughout the whole altitude of the prism. Moreover, those different waxes are resplendent with the most vivid colours, and are joined together with the most consummate art, and almost fused together in such a way that you would suppose you were seeing the discoloured veins of smooth marble. Certain parts of the picture seem to be less perfect, e.g. one of the arms, in which the plates of those colours between the most intense light and shade are missing; with them the transition from the one to the other would have been gentler and more even. But I imagine that this was due not so much to a defect in the art, as to the artist's judgment, since, having exerted himself to give an excellent demonstration, in Erigone's hair and entire head, of his admirable skill in mixing well those

* See Günther's pamphlet on the art of painting with dry pigments ([*Praktische Anweisung zur*] *Pastel-Farben*), which he published in [Nuremberg in] 1762, p. 130[–35].

boundaries of light and shade, he seems not to have been so concerned about the remaining details.[296]

These are the most general features which I was able to observe. It is probable that the entire picture was composed from those less elaborate parts in the manner of an artistic painting. However, they could by no means have penetrated the base of the prism, smeared with wax, by a movement parallel to the axis (as some idly believe), with pigments watered down by some unknown liquid; nor could they have filled it up entirely with their own remains and with that picture. If Mayer actually said that this was the case, as I have heard some people claim, I imagine that these were not the words of an inventor revealing the secrets of his art to a friend, but of an inventor jocularly concealing these secrets from those who showed an untimely curiosity and who were unskilled. It may readily be conjectured from this that this picture cannot resemble the usual artistic work in all particulars, because wax can be liquified, bent, and compressed, whereas marble can not. I do not undertake here to follow up the conjectured profits of the work, which without doubt were great for the inventor; I merely add that certain parts were connected up in such a way that it seems impossible for them to have been joined together unless with liquid wax injected, as it were, in the tubes and fissures, where others already well congealed had been hollowed out again.

In thinking about his chromatometer, Mayer could have alighted upon this type of painting in two ways; either by creating for himself in rough outline a pyramid or prism, from which all the triangles could be cut, similar to that which we created for ourselves above; or, which seems more probable, thinking about chromatometers *for himself* which would be *reliable*, *durable* and *readily reproducible*, such as they ought to be, if faithful and easily made likenesses of the touchstone are to be obtained. For first of all, pigments set in wax are more vivid and more durable; then, once a mixture has been made, hexagonal prisms could be completed in the way in which they are usually made of wax. Those little squares could be cut from them individually, or else, if they were all joined together in one prism, a complete chequered triangle with a jointed section could be produced.

REMARKS ON THE SELENOGRAPHIC TABLE

It gives me pleasure to be transmitting to the learned world Mayer's selenographic table; not only because I am obliged to, but also since the distinguished author [Mayer] entrusted it to his relatives at the end of his life and expressly requested that it be made public. As for the original itself, I have taken care that it should be engraved by the local artist Kaltenhofer, whose wonderful accuracy in drawing is praised no less than that of Mayer in observing by those skilled in art, [who] assuredly prefer the reliable and not inelegant engraving of this artist to the more elegant but less reliable [engraving] of others.[297] And so there can be no doubt at all about the correspondence between the copies etched in copper plate and the original; and I have no word of caution to utter except that on account of contraction of the map the lunar axis and the diameter of the equator are unequal. In most printed copies of the common double-sized map, I have found that the ratio of the former to the latter is $=99:98$. Thus care will have to be taken in comparing the distances of the [lunar] markings from the equator with the [Moon's] semidiameter—the sine of the latitudes, as it were, with the sine of the whole [i.e. sin 90°, or 1]—in investigating their latitudes. The magnitude of the errors which can arise from this, especially if the markings are circumpolar, is clear without any warning from me. Before I list the few observations which I myself have added, I shall supply certain details relating to the history of this selenographic table from two of Mayer's writings. [This is] for the sake of those who are not skilled in the German tongue in which they are written, or who have never happened to inspect them inasmuch as they are not accessible everywhere. [I am] leaving the rest, the greatest part [of this study], to the reader, for whom Mayer bequeathed nothing on this subject which he might look at apart from the pair [of treatises] in my care. The one of these is the account of the axial rotation of the Moon and the apparent movement of lunar markings, included in Vol. 1 of the writings of the Cosmographic Society;[298] and the other, a report of his about the artificial lunar globes, published at Nuremberg in 1750.[299] The reader who is desirous of knowing more than there is room for here is referred to these works.

Mayer was around 25 years old when he proposed, at the beginning of 1748, to represent the lunar eclipse which occurred on 8 August of the same year;[300] and at the same time to predict accurately from it the times at which the individual markings passed into the Earth's shadow, and [the times] of their emergence from it. He at once discovered how many errors there were in the selenographic tables of Hevel and Riccioli. So, from that time onward, he directed his attention to that aspect of astronomy which, if not entirely neglected by astronomers hitherto, was certainly not treated with the same diligence that they expended in the perfection of other fields. He improved selenography and left it to posterity in a much more complete state than he had received it from his predecessors. He maintained this [effort] for around the space of 18 months by continuing, until the middle of 1749, observations begun in February 1748. For these observations he used telescopic tubes no less than 9 feet and no greater than 12 [in focal length], choosing these two limits so that the lunar disc might be completely in the field of view, and magnified considerably. To begin with he used the common cross-wire arrangement [in the eyepiece of his telescopes] consisting of four threads intersecting at an angle of 45°; but subsequently changing his plan (in accordance with his usual practice), he discovered for himself [a micrometer] which he always used afterwards.*

Being provided with these instruments, he [Mayer] made a great number of very precise observations. With the help of these, when he had illustrated in an outstanding manner the most difficult part of selenography, i.e. the theory of libration, he applied himself to producing a more accurate diagram of the lunar surface than had hitherto appeared. With a view, at the same time, to conferring a great benefit on the whole of selenography, he proposed to construct artificial lunar globes, such as Hevel had judged to be already completed. He has amply set out their use and nature in the publication mentioned above.[304] Thus, in the first place, he brought markings whose positions he had determined from observations by calculations based upon the formulae which he had bequeathed in his treatise on the [axial] rotation of the moon into a network for drawing a lunar map whose diameter is 15 inches—these [surface features] being the foundation stones, as it were, for other conclusions. Then, relying on his visual judgment, he sketched the positions of other spots and the shapes of seas and canals; with such care and circumspection,

* Both the construction and the use of this micrometer he himself expounded in the writings of the Cosmographic Society mentioned above.[301] It consisted of several parallel lines drawn in Indian ink on a glass disc with outstanding skill. In the tube employed by Mayer, the gap between a pair of these [parallel rulings] corresponded to approximately one minute of a great circle arc. He investigated their individual values by repeated experiments and in varying ways, and constructed for himself a table of the values thus derived. Another three parallel lines, the middle one of which passed through the centre of the disc, cut these parallels at right angles; their spacings were [also] equal, but more or less ten times greater than the first-mentioned rulings. Micrometers of this type have now been made with great skill by Brander at Augsburg with the finest of lines cut into glass discs. See Kästner's Astronomical treatise Vol. 2, Treatise 7[302] and the pamphlet of Lambert's, which he published at Augsburg 1769, dealing with Brander's micrometers.[303]

however, that he usually spent three or four hours drawing a region of the Moon's surface scarcely equal to a hundredth part of the whole orb. There are 40 exceedingly fine drawings of this kind. From these sketches, when they had at length been duly assembled, there came into being the greater lunar map.[305] A copper engraving of a smaller copy, most skilfully drawn by the author himself, has been appended to this work.[306] From that larger drawing—divided into 12 parts of which six constitute the northern part of the Moon and six the southern part—he has now skilfully drawn eight parts unfolded according to the rules of projection, so that the globes may be supposed from these to be around 15 inches [in diameter]. Thus four parts, two northern and two southern, have still to be drawn. Indeed, [so does] that part of the orb which lies between the western limb of the Moon and the meridian passing through the [selenographic] longitude 30° West. [This applies], moreover, to those regions on the lunar surface that vanish from our sight at the time of mean libration, in accordance with which the shape of the Moon is displayed on these maps. At present, only six of the segments mentioned exist etched in copper plate.[307] Perhaps it will be granted to readers of posterity to make a welcome discovery concerning the progress of this work.[308] In the meantime, regarding this delay, it is some consolation (although a poor one) that this work was not interrupted by Mayer's death. That most learned man, being intent on other discoveries and also for reasons which it is of little importance to relate here, laid it aside a long time before his death; and indeed so completely did he do so, as I was told by his friends, that he seemed scarcely to be able to bear being asked by anyone about the lunar globes.[309]

The justification for the purpose and use of the lunar globes, which is especially obvious in selenography, and is not restricted to displaying the face of the full Moon according to the various effects of libration, is apparent when one considers why the map which is presented to astronomers here, appears in a form so different from the usual one. For it can readily be seen that the artificial lunar globe, which can serve to represent all the libratory phases of the Moon, and what's more, to solve various problems concerning the positions of the markings, must constitute not so much an exact pictorial image of the lunar hemisphere illuminated from an infinitely distant shining point, as a representation of the individual parts seen under the most suitable light and duly combined. Thus, in order to obtain this result, whatever lunar region Mayer drew at that time, he executed so that it might approximate to both light and shade in the neighbouring phases, or at least be not very different from them; because at that time, on account of the notable size of the shadows, the lower-lying places scarcely disappear from the observer's sight.

What I have said concerning the globe holds true for the most part as regards the selenographic table, since it must be considered as an orthographic projection of the artificial globe on to the plane of its meridian, which seems to us to coincide in mean libration with the Moon's disc. Therefore the Moon never presents that appearance to us, but it would appear [so] if we were permitted to look simultaneously at exposed objects on its surface in a light particularly suited to enabling them to be readily distinguished.

And so Mayer strove to render symbolically the shadow lying near the high regions, since it could not be more fully illustrated than by this pictorial method (being undoubtedly of such a kind as agrees with objects arranged on a spherical body). For in the segments mentioned above, [the depth of shading] is about equal to eight times the altitude of the object. He seems to have followed these methods as far as he could in the maps as well.

This selenographic map is therefore of little use in observations of lunar eclipses, since in certain regions, especially southern ones, it differs from the appearance of the full Moon to such an extent that it seems to be a picture of some region as yet unseen by us rather than of the same one whose only difference is the shadows. Nevertheless, on account of its containing many precise determinations for many of the markings from which one is free to construct a picture of the full Moon with little difficulty and with those not very difficult pioneer observations, I am sure that it will be a very welcome future gift to astronomers. [This is] to say nothing of the pleasure which a picture of our Earth's satellite will bring to lovers of natural philosophy, so accurately depicted is it, with such consummate skill; with it they can carry out their own calculations more easily than if they were looking at the Moon itself, and either construct conjectures, hypotheses and systems from it or destroy those built up by others.

But I now proceed to tell of the few details which I have added. I have given numbers to the intersections of the meridians with the equator, and of the parallels with the actual circumference of the disc or from the meridian straight to the primary meridian. With the help of these and of the catalogue which contains their selenographic longitudes and latitudes, the more important features can easily be located. In this I have considered that I should follow the renowned Lambert, who added similar numbers to his own figure of the full Moon appended to the Berlin Ephemerides for the year 1776.[310] For I was unwilling to obscure the picture by giving numbers to the spots and, even if I had wished it, I could not have done so without great loss of clarity. Concerning the Catalogue the following points are to be noted. The longitudes and latitudes of the markings whose names have been expressed in italics, to which the number in the fifth column relates, have been calculated by Mayer from his own observations, and must be considered as [providing] the basic framework of those sketches: they were taken by me from his diary. I duly determined the positions of the remainder both from the larger map and from the segments themselves, with a pair of compasses and by calculation. Under the selenographic position of the spots we understand the position of their centres and also of the rills which are generally present, provided that there is only one rill of this kind and that [it is] near to the centre of the spot. In most cases a little point could be seen made by Mayer on the map with the top of the compasses. The + sign is prefixed to the western longitudes and northern latitudes, and the − sign to the eastern longitudes and southern latitudes, as the signs placed in the margin of the figure show.

I could easily have made this catalogue much more comprehensive, but I was afraid that, in seeking Riccioli's names for the marking in this figure, I might in the meantime have made too many mistakes, especially in those near the limb. Not all the discrepancies which the careful reader will observe between the position of a marking indicated in the catalogue and that which it occupies in the adjoining figure, are to be put down to my calculations; for, after the task was almost finished, I saw that not only did the first original differ from the smaller copy, but also that both differed from the segments and these again from the positions defined by observation and computation. On the other hand, since the difference is never so great as to be able to cause doubt to anyone seeking a marking in the figure by following the numbers prescribed in the catalogue, I have not changed anything, especially since astronomers will be thankful to know of this variation. I have indicated the more noteworthy differences.

BIBLIOGRAPHICAL REFERENCES

Introduction

1. An entry in the Esslingen family register refers to the birth of a son (Johannes Wolfgang) to the Mayers on 13 December 1724, and another to this baby's death only six days later. The family's transfer from Marbach would therefore appear to have taken place some months earlier than Mayer recalls in his autobiographical account of his childhood years published by Baron Franz Xavier von Zach, *Monatliche Correspondenz zur Beforderung der Erd- und Himmels-Kunde*, ix (1804), 415–432.

2. Mayer's father died on 12 August 1731 at the age of 48, after an illness lasting several months.

3. Christian, Georg Wilhelm, Margaretha, Justina, Eva Catherina, (Johann) Tobias; the last two being the surviving offspring of Mayer's father's second marriage, to Maria Catherina Finken.

4. Documentary evidence to support this belief has been published by Paul Eberhardt, 'Urkundliche Beiträge zu der Jugendgeschichte des Astronomen Johann Tobias Mayer', *Literarische Beiträge des Staatsanzeigers*, 1908, pp. 177–187; reprinted with improvements in: *Aus Alt-Esslingen* (Esslingen, 1924), pp. 207–224.

5. In the Minutes of the Esslingen Society of Antiquaries for 6 March 1925, one reads: 'Dr Haffner teilt mit, dass er im Archiv verschiedene Ansichten vom Alten Spital gefunden habe, die von Tobias Mayer gefertigt seien, der erst als Fundenkind in dem Spital aufgenommen war. Er war damals knapp 14 Jahre alt. Der Wert der Bilder liege darin, dass sie, wenn auch technisch schülerhaft ausgeführt, doch die Art der ganzen Anlage des Gebäudekomplexes zeigen.'

6. The relevant extract from the Esslingen town-council Minutes of 16 December 1738 is: 'G. Geiger, älterer Feuerwerker, gibt per Memoriale geh. zu erkennen, dass er dem in dem Fundenhaus sich befindl. Tobias Mayer wegen seiner zur architectur bezeugten ausserordentl. Triebe die affection erzeigt, denselben in der Artillerie Geometrie auch Planimetrie und Fortification mit solchem Fleiss zu unterrichten, dass diesem jungen Menschen nichts mehreres mangle, als zu seinem Fortkommen ein noch auszufertigen seyender Lehr-Brief. Wie nun denen Fundenkindern ein Lehrgeld sonsten pflege gereicht zu werden; als bittet Supplicant vor die mitgetheilte Information ein ergibiges Lehrgeld gnädig zu adjudiciren; anerwogen die allbereits schon vormahls resolvirte 3 fl. kaum sufficient vor die Farben, welche gebraucht worden.'

7. See *Catalogus Praelectionum publice et privatim in Academia Georgia Augusta*, 1751, 1755–1761 (incl.). A lengthy treatise entitled 'Artillerie' is contained among Mayer's still unpublished manuscripts in Göttingen (cf. 46).

8. Mayer was taught Latin, Greek, logic, rhetoric, geography, and history. For further information, the reader should consult: D. Mayer, 'Die lateinische Lehranstalt Esslingens vor hundert Jahren und seit hundert Jahren', *Programme des Gymnasiums in Esslingen* (Esslingen, 1900).

9. The main sources of information about Mayer's youth in Esslingen can be found for example, in: D. Hausleutner, *Schwäbisches Archiv* (1793), pp. 385–392; J. J. Keller, *Beschreibung der Reichsstadt Esslingen und ihres Gebietes* (Esslingen, 1798); von Zach, *loc. cit.* I, viii (1803), pp. 257–270 and ix (1804), pp. 45–56, 415–432; Benzenberg (*op. cit.* 11); C. F. Ofterdinger, *Mathematisch-naturwissenschaftliche Mitteilungen*, ii (1887), pp. 116–132.

10. The full title, in German, is: *Neue und allgemeine Art, alle Aufgaben aus der Geometrie vermittelst der geometrischen Linien leichte aufzulösen; insbesondere wie alle reguläre und irreguläre Vielecke, davon ein Verhältnis ihrer Seiten gegeben, in den Circul geometrisch sollen eingeschrieben werden, sammt einer hiezu nötigen Buchstaben-Rechnenkunst und Geometrie.*

11. This work (cf. 10) was reprinted with a 71-page introduction, by J. F. Benzenberg under the title: *Erstlinge von Tobias Mayer, aufs neue herausgegeben, nebst einigen Nachrichten von seinen Erfindungen und seinem Leben.* (Düsseldorf, 1812.)

12. A description of it is given by Mayer in: *Kosmographische Sammlungen auf das Jahr 1748* . . . (Nuremberg, 1750), 1–11; and reprinted by Johann Heinrich Lambert in: *Anmerkungen über die Branderischen Mikrometer von Glase* . . . (Augsburg, 1769), pp. 31–48.

13. The previous information in this paragraph is based upon Mayer's own remarks in his lengthy treatise on the Moon's libration (cf. 14, p. 93).

14. 'Abhandlung über die Umwalzung des Mondes um seine Axe, und die scheinbare Bewegung der Mondsflecken, worinnen der Grund einer verbesserten Mondsbeschreibung aus neuen Beobachtungen gelegt wird,' in: *Kosmographische Sammlungen auf das Jahr 1748* ... (cf. ref. 12), pp. 52–183. It is from this treatise that the brief description of Mayer's observing technique given in the text has been abstracted.

15. On this subject, see J. Delambre, *Histoire de l'Astronomie au Dix-Huitième Siècle* (Paris, 1827), pp. 429–449; and J. Browne's article on Tobias Mayer in the 7th edition of the *Encyclopaedia Britannica* (Edinburgh, 1842), which was almost entirely based upon the former source.

16. 'Bericht von den Mondskugeln, welche bey der kosmographischen Gesellschaft in Nürnberg, aus neuen Beobachtungen verfertiget werden durch Tobias Mayern, Mitglied derselben Gesellschaft. Zu finden in der Homannischen Officin.' (Nuremberg, 1750.) Mayer is usually credited with having written this 24-page pamphlet; it may, however, have been composed by Johann Michael Franz, the Director of the Homann Cartographic Bureau.

17. One reason was that Mayer left Nuremberg early in 1751; another, that the Homann firm went bankrupt as a result of the over-ambitious projects initiated around this time by Franz. See S. Ruge, 'Aus der Sturm- und Drang-Periode der Geographie', *Zeitschrift für Wissenschaftliche Geographie*, v (1885), pp. 249–260; also the article by Christian Sandler (cf. 20).

18. D. B. Herrmann, 'Georg Christoph Lichtenberg und die Mondkarte von Tobias Mayer,' *Mitteilungen der Archenhold-Sternwarte Berlin-Treptow Nr. 72*, 1965.

19. Abraham Gotthelf Kästner, Elogium Tobiae Mayeri (Göttingen, 1762). The original Latin text is as follows: '—praxium mathematicarum usus aliis impertiit, emolumentum inde ad se redundare sensit, quoniam vulgus scientias ut amicitias utilitate probat. Hic substituri erant multi, qui practices mathematicos se profitentur, Mayerus vidit, praxim quae vulgari theoria utcunque firmatur, profundiore, liberandam esse a naevis & perficiendam. Hinc, & artem inveniendi mathematicam, quantamcunque nostro aevo reddiderunt summa ingenia, animo est complexus, & feliciter ad praxin transtulit.'

20. Between 1747 and 1754, Mayer made no fewer than 31 maps which bear his name. C. Sandler, 'Die Homännischen Erben,' *loc. cit.* 17, vii (1890), pp. 333–355 and 418–448, states his belief on p. 437 that Mayer made in all 40 maps, mainly of small regions.

21. For background information on Göttingen and its university, see, for example, A. Saathoff, *Geschichte der Stadt Göttingen seit der Gründung der Universität* (Göttingen, 1940). Chapters 3 and 4, pp. 26–63 are the most relevant to the present study.

22. Unpublished letter from Gerhardt Adolph von Münchhausen in Hanover to Tobias Mayer, Mathematician in Nuremberg, dated 26 November 1750, preserved in the University archives in Göttingen under: *Personalakte Tobias Mayer* Nummer 4/Vb 18.

23. In the records of the *Evangelisches Stadtpfarramt* in Neckarbischofsheim one finds that Johann Christoph Gnüge came originally from Gotha (Thüringen) and began working in that parish in 1704 as second-charge and from 1705 as first-charge of the ministry. Maria Victoria was the youngest of six children, and was born on 30 July 1731.

24. The names and dates of birth of Mayer's eight children, as recorded in the 'Kirchenbuch der Albani Gemeinde' at the *Evangelische Kirchenbuchamt* in Göttingen, are as follows: Johann Tobias (5 May 1752), Elizabeth Clara (16 January 1754), Georg Moritz (20 March 1755), Julius Georg (23 August 1756), Johann Friderich (7 November 1757), Juliana Sophia Maria (10 February 1759), Johann Christoph (26 March 1760), and Sophia Elizabeth Victoria (20 May 1761).

25. Most of this correspondence was published by Y. K. Kopelevich, *Istoriko–Astronomijezki Issledowanje*, v (1959), pp. 279–427; and the rest (four letters from Euler to Mayer) by Y. K. Kopelevich and E. Forbes, *ibid.*, x (1969), pp. 285–310.

26. Kopelevich, *ibid.*, pp. 413–414: Mayer to Euler, 6 October 1754. Further information regarding the concessions granted to Mayer, and his decision to remain in Göttingen, is contained in three letters in his *Personalakte* 4/Vb 18, pp. 15–17.

27. The letter from Maupertuis, dated 27 August 1754, is still unpublished and preserved under: *Cod. MS. philos. 159* 'Briefe von und an J. Tobias Mayer', at the Göttingen University library.

28. 'Novae tabulae motuum solis et lunae,' *Commentarii Societatis Regiae Scientiarum Gottingensis*, ii (1753), pp. 383–430.

29. Alexis Clairaut, *Théorie de la lune déduite d'un seul principe de l'attraction* . . . (Berlin 1752). See also Y. K. Kopelevich, 'The Petersburg Astronomy Contest in 1751', *Soviet Astr.—AJ*, ix (1966), pp. 653–660.

30. Kopelevich, *op. cit.* 25, pp. 391–393.

31. 'Tabularium lunarium in Commentt. S.R. Tom. II contentarum usus in investiganda longitudine maris,' *loc. cit.* 28, iii (1754), pp. 375–396.

32. The bulk of the correspondence which ensued as a result of Michaelis's initial letter on 4 August 1754, is preserved in the Staatsarchiv Hannover; *Hannover Des. 92* XXXIV No. II, 4, a^1, and entitled: 'Betr. der von Seiten des Prof. Tobias Mayer in Göttingen gelöste englische Preisfrage über die Bestimmung der Longitudo maris, 1754–1765.' A number of important letters from Best to Michaelis and von Münchhausen during this same period are contained in *Cod. MS. Michael. 320*, pp. 535–671 at the Göttingen University library.

33. A brief survey of the work of the Board of Longitude is to be found in: E. G. Forbes, 'The Board of Longitude Papers at the Royal Greenwich Observatory', *Actes du XIe Congrès International d'Histoire des Sciences*, iii (1968), pp. 103–107.

34. Bradley's two letters to the Admiralty Secretary (Cleveland), dated 10 February 1756 and 14 April 1760, are printed in *op. cit.* 35, pp. cix–cxv; and again in S. P. Rigaud, *Miscellaneous Works and Correspondence of the Rev. James Bradley, D.D., F.R.S.* (Oxford, 1832), pp. 84–89.

35. The first diagram and description of Mayer's Repeating Circle was not published until 1770, when it appeared in N. Maskelyne's edition of the *Tabulae motuum solis et lunae novae et correctae auctore Tob. Mayer: quibus accedit methodus longitudinum promota eodem auctore* (London, 1770). Cf. ref. 74.

36. 'Betr. den Contract wegen eines in London zu verfertigenen Quadranten zum Gebrauch des Observatorii astronomici 1754–1755,' in Mayer's *Personalakte* 4/V f/2.

37. 'Observationes astronomicae A. 1753 Gottingae habitae', *loc. cit.* 28, iii (1754), pp. 441–454.

38. See Treatise II: 'Observationes astronomicae quadrante murali habitae in observatorio Gottingensi', the content of which is the justification for the statements contained in the preceding three sentences of this paragraph.

39. 'Instrumenti goniometrici, quod astrolabium vocant, structura emendatior,' *Göttingische Anzeigen von gelehrten Sachen* (1759), p. 995. Mayer raised the status of the long-known astrolabe to that of a precision instrument by introducing a micrometer screw, a telescope with a cross-wire, and a water-level. See E. G. Forbes, 'Tobias Mayer's New Astrolabe (1759)—its principles and construction', *Ann. Sci.*, xxvii (1971), 109–116.

40. 'Colorum ex pigmentis commixtis oriundum computatio,' *ibid.* (1758), pp. 1385–1389. This is an abstract of the contents of the fourth treatise in this volume.

41. 'Theoria magnetica,' *ibid.* (1760), p. 633; and 'Computus declinationum ex inclinationum magneticarum, ex theoria nuper exhibita deductus', *ibid.* (1762), pp. 377–379. These are abstracts of the treatises cited in ref. 46 (below).

42. cf. *op. cit.* 7.

43. Sophus Ruge (*op. cit.* 17) and Christian Sandler (*op. cit.* 20) both provide information regarding this transfer, and the misfortunes that were destined to accompany it. More manuscript material on this subject, which was not used by either of these authors in preparing their respective articles, is preserved in the Göttingen University Archives under: 'Univ. Gött. Philos. Facultät Lehrer' Ph 1/18 4Vb, Nr. 23, 27.

44. 'Eigenschaften und Vorzüge der neuen Kunst, Gemählde mit Farben zu drucken.' This unpublished document, in which Mayer lists eight advantages of his discovery, is contained in his *Personalakte* (cf. ref. 22), pp. 33–36. See E. G. Forbes, 'Tobias Mayer's theory of colour-mixing and its application to artistic reproductions', *Ann. Sci.*, xxvi (1970), 95–114.

45. It was apparently entitled 'Artis qua picturae datae ectypa multiplicantur specimen exhibitum' (cf. *loc. cit.* 39, p. 402). This art is described by Mayer's eldest son, also christened Johann Tobias, in a letter to Johann Heinrich Lambert dated 2 November 1774, published in J. Bernouille, *Johann Heinrich Lamberts deutscher gelehrter Briefwechsel*, ii (Berlin, 1782), pp. 452–454.

46. Catalogued as item 15_{18} in 'Tobias Mayers Nachlass, aufbewahrt in der K. Sternwarte', *Verzeichniss der Handschriften im Preussischen Staate*, i (Berlin, 1894), pp. 154–158.

47. A. Coulomb published three memoirs 'sur l'électricité et le magnétisme,' in *Histoire de L'Académie Royale des Sciences Année MDCCLXXV* (Paris, 1788): 'Premier Mémoire', pp. 569–577; 'Sécond Mémoire', pp. 578–611; 'Troisième Mémoire', pp. 612–638. The second memoir is that which contains his demonstration of the inverse-square law of magnetic attraction and repulsion.

48. *Op. cit.* 46, Caput 2, §12. The original Latin text is: 'Explicantur phaenomena gravitatis universalis, elasticitatis, duritiei corporum, etsi causa earum mysterium sit, neque melius faciliusque explicarentur vel ipsa verissima causa patefacta. Quin ob id ipsum, quod non valde utilis sit causarum remotiorum cognitio, natura eas abscondidisse videtur, adeo ut non possimus non est in hoc veritatem agnoscere illius dicti; Naturam nihil frustra facere.'

49. *Georg Christoph Lichtenbergs Vermischte Schriften. Neue vermehrte, von dessen Söhnen veranstaltete Original-Ausgabe. Erster Band.* (Göttingen, 1844.) cf. p. 290.

50. *Ibid.*, p. 217. The original statement is, of course, in German.

51. 'Betr.: den Hauskauf des Profess. Philosoph. Mayer 1754–1757.' Cf. 22, pp. 1–10.

52. *Hannover Des. 92* XXXIV No. II, 4, a^1 (cf. 32), p. 55.

53. 'Verhandlungen betr. das Vermächtnis Tobias Mayers, 1762–96,' designated Etat 32, 1 in the archives of the Akademie der Wissenschaften, Göttingen (cf. p. 1).

54. *Ibid.*, p. 2b, records all seven questions put to Mayer on his deathbed, and his replies.

55. *Ibid.*, p. 3b.

56. *Loc. cit.* 52.

57. Letter from Best to Michaelis; London, 2 April 1762. P. 643 in *Cod. MS. Michael. 320* at the Göttingen University library.

58. *Loc. cit.* 22, pp. 25, 26.

59. The foregoing facts in this paragraph have been abstracted from the section of the correspondence *loc. cit.* 52, pp. 12–39.

60. *Royal Greenwich Observatory* (R.G.O.) MSS. 'Board of Longitude', (P.R.O. Ref. 533), cf. Minutes of Meeting on 5 September 1763.

61. Cf. E. G. Forbes, 'Tobias Mayer's Lunar Tables', *Ann. Sci.* xxii (1966), pp. 105–116.

62. *Loc. cit.* 52, pp. 18–20.

63. These had apparently been made at the Nuremberg School of Art, under supervision of the Director Georg Martin Preissler (1700–1754) and his successor. See *loc. cit.* 46, item Mayer 15_{52}.

64. Johann Tobias Mayer to Johann Heinrich Lambert, 18 February 1772, in Bernoulli, *op. cit.* 45, p. 432.

65. P.M. from Balck (Hanover) to Michaelis, 17 October 1763, designated Scient. 100 Nr. 1 in the archives of the Akademie der Wissenschaften, Göttingen.

66. 'Betr. den Ankauf der Kupferstiche behuf einer Monds Charte von der verwitweten Profess: Mayern. 1763,' in Mayer's *Personalakte* (cf. 22).

67. *Loc. cit.* 52, p. 8.

68. His letter, dated 1 November 1764, is contained in *ibid.*, p. 11.

69. E. G. Forbes, 'The Bicentenary of the Nautical Almanac (1767),' *Br. J. Hist. Sci.*, iii (1967), pp. 393–394.

70. E. G. Forbes, 'The Origin and Development of the Marine Chronometer', *Ann. Sci.* xxii (1966), pp. 1–25, contains an account of the circumstances connected with Harrison's claim.

71. This is stated by Maskelyne in his memorial to the Board of Longitude on 9 February 1765. Cf. *loc. cit.* 60, p. 41.
72. W. E. May, 'Early Reflecting Instruments,' *Naut. Mag.* No. 145 (1945) pp. 21–26.
73. T. Mayer, *Theoria Lunae juxta Systema Newtonianum* (London, 1767).
74. The English title was: 'New and Correct Tables of the Motions of the Sun and Moon, by Tobias Mayer: to which is added the Method of finding the Longitude improved; by the same Author'. Cf. ref. 35.
75. A very full biography of Lichtenberg's life, up to the year 1775 which happens to be that in which the *Opera inedita Tobiae Mayeri* was published, is to be found in: Otto Deneke, *Lichtenberg erzählt sein Leben* (Munich, 1944).
76. In G. von Selle, *Die Matrikel der Georg-August Universität zu Göttingen 1734–1837* (Hildesheim and Leipzig, 1937) one finds that 203 students enrolled for the first semester in 1763 as compared with 15 and 47 for the years 1761 and 1762 respectively.
77. M. L. Mare and W. H. Quarrell, *Lichtenberg's Visits to England as described in his Letters and Diaries* (Oxford, 1938).
78. *Loc. cit.* 39 (1767), pp. 401–403; *ibid.* (1769), pp. 665–667.
79. *Loc. cit.* 76, p. 108.
80. More details concerning Lichtenberg's geodetic work are to be found in: Paul Hahn, *Georg Christoph Lichtenberg und die exakten Wissenschaften* (Göttingen, 1927), pp. 16–20.
81. G. C. Lichtenberg, 'Observationes Astronomicae per Annum 1772. Et 1773. Ad Situm Hannoverae, Osnabrugi et Stadae Determinandum Institutae,' *Novi Commentarii Societatis Regiae Scientiarum Gottingensis*, vii (1776), pp. 210–232.
82. S. Günther, 'G. C. Lichtenberg und die Geophysik,' *Abhandlungen der K. K. Geographischen Gesellschaft in Wien*, i (1889), pp. 119–135, tabulates the two independent sets of results on p. 123 of his article as follows:

	Northern Latitude		*Eastern Longitude*	
	(Lichtenberg)	(Gauss–Schumacher)	(Lichtenberg)	(Gauss–Schumacher)
Hanover	52° 22′ 18″	52° 22′ 25″	$0^h\ 30^m\ 28^s$	$0^h\ 29^m\ 36^s$
Stade	53 36 5	53 36 8	0 28 8	0 28 34
Osnabrück	52 16 14	52 40 48	0 21 44	0 22 29

The longitude is based upon the Paris meridian: $8^m\ 31^s$ must be added to the times quoted in this table if the Greenwich meridian were chosen as the standard of reference.
83. The respective magnetic variations for Hanover, Stade, and Osnabrück in 1773 were found to be 17° 20′, 17° 18′, 17° 25′ East of North (cf. *ibid.*).
84. Translated from Hahn, *op. cit.* 80, p. 19.
85. 'Nova methodus perficiendi instrumenta geometrica et novum instrumentum goniometricum', *loc. cit.* 28, pp. 325–336.
86. cf. 35.
87. cf. 65.
88. P.M. from Lichtenberg to some official [perhaps Christian Schernhagen] in the Hanoverian Government, dated 2 March 1773, in *loc. cit.* 22, p. 28. Lichtenberg distorts the true situation somewhat when writing to Lambert from Stade on 14 September of that year (*loc. cit.* 45, pp. 457–458); for he says that he had broached the subject with some important people when in Hanover that summer, with the result that he had been *asked* three months ago (i.e. in *June*) to submit a written notice about it.
89. Just when he had received this assurance, it is hard to say. He might have met Maskelyne during his visit to England in 1770, and since the latter had just finished editing Mayer's solar and lunar tables (cf. *loc. cit.* 35) this would have been a natural topic of conversation. On the other hand, Mayer's eldest son (Johann Tobias) had brought the matter directly to Maskelyne's attention in 1772 by asking a friend in Berlin to write to another friend in London who would speak to Maskelyne about it, and by writing himself to Maskelyne (cf. *loc. cit.* 45, pp. 438–439; 444; and 448–449).
90. Cf. *loc. cit.* 41, 40, 98 for references to these respective abstracts.
91. 'Von der hann. Regierung mit der Herausgabe nachgelassener Schriften Tob.

Mayer's betraut, 1773,' *Scient.* 100 Nr. 2 in the Akademie der Wissenschaften, Göttingen (cf. *loc. cit.* 65).

92. Lichtenberg to Lambert; Stade, 14 September 1773, *loc. cit.* 45, p. 458.

93. Lichtenberg to Dieterich; Stade, 28 June 1773, in: A. Leitzmann and C. Schüddekopf, *Georg Christoph Lichtenberg: Briefe* (Hildesheim, 1966). Letter 78, Vol. I, p. 142.

94. Lichtenberg to Kaltenhofer; Stade, 23 August 1773. *Ibid.* Letter 87, p. 162.

95. Lichtenberg to Schröter; Göttingen, 21 April 1791. *Op. cit.* 93. Letter 589, Vol. III, p. 23.

96. Lichtenberg to Kaltenhofer; Stade, 14 June 1773. *Op. cit.* 93. Letter 74, Vol. I, p. 135.

97. See ref. 92.

98. 'Versuch einer Erklärung des Erdbebens,' *Hannoverische nützliche Sammlungen* (1756) §19, pp. 290–296. This article is *not* listed by J. C. Poggendorff, *Biographisch-Literarisches. Handwörterbuch*, ii (Leipzig, 1863), p. 91. An abstract of Mayer's explanation appears in *loc. cit.* 39, pp. 314–316.

99. The relevant extract is to be found under the more general heading: 'Vorschläge zu Themen der mathematischen Hauptpreisschrift, u.s.w.,' in *Scient.* 182′ at the Akademie der Wissenschaften, Göttingen. Lichtenberg remarks (*inter alia*) that Mayer's explanation was either intended as a joke, or must certainly be regarded as being one.

100. Translated from a quote by Hahn, *op. cit.* 80, p. 23.

101. The following entry on p. 258 of Lichtenberg's third wastebook, written in June 1773, is very revealing in this context: 'I have always thought that I might yet make some great invention; to add a small touch to other people's works or to ward off the flies I have never thought worth my while'. (Quoted by Stern, *op. cit.* 122, p. 49.)

102. Compare the letters from Lichtenberg to Schernhagen; Stade, 28 September 1773, in the former's *Nachlass* (Kasten I) in the Göttingen University library; and from Lichtenberg to Dieterich (*loc. cit.* 93, Letter 95, Vol. I, p. 177), which was written at about the same time.

103. Letters 102 (pp. 186 ff.) and 105 (pp. 191 ff.) respectively, in *loc. cit.* 93.

104. Hahn, *op. cit.* 80, p. 23.

105. P. Requaedt, *Lichtenberg* (Stuttgart, 1964), Ch. III, p. 82, regards Lichtenberg's approach to all problems of investigation as being essentially a combination of his search for subjective truth through self-observation and his search for objective truth through the observation of nature.

106. 'Betr. die Bestellung des Professors der Philosophie Georg Christoph Lichtenberg 1770–1795,' in Lichtenberg's *Personalakte* Ph 1/37 4Vb No. 39 in the University Archives, Göttingen. This document contains the sole reference, in all these papers, to his involvement in the editing of Mayer's unpublished writings.

107. *Op. cit.* 77, p. 52.

108. This supposition would appear to be borne out by the fact that in a postscript to his letter to Ernst Gottfried Baldinger, written on 24 January 1775, Lichtenberg remarks that he had sent the preface to [Christian Gottlob] Heyne—a well-known classical scholar who was also a professor at Göttingen—who had altered only a single word on the second page. *Op. cit.* 77, p. 70.

109. *Op. cit.* 15, p. 447.

110. Mayer supplies his own proof of the cosine formula for a spherical triangle in Part II of his unpublished treatise 'Die Regeln der sphaer. Trigonometrie aus einem einzigen grunde hergeleitet.' (*loc. cit.* 46, item 15_{10}.)

111. G. van Biesbroeck, 'The Astronomical Works of Olaus Roemer,' *J. R. astr. Soc. Can.*, vii (1913), pp. 342–358. This was actually translated from the original French version entitled 'Les Travaux Astronomiques D'Olaüs Roemer', *Ciel Terre* xxxiv (1913), pp. 152–167.

112. A detailed description and diagram of this meridian transit instrument are to be found in: P. Horrebowii, *Operum mathematico-physicorum*, iii (Copenhagen, 1741).

113. Elis Strömgren, 'Om Ole Rømers Meridianobservationer Til Bestemmelse af Fikstjerners Rektascensioner samt om Formeln til Saadanne Observationers Korrigering for Fejl i Meridian-Instrumentets Opstillung,' *Nord. astr. Tidsskr.* xvii (1936), pp. 17–26.

114. T. Eibe and K. Meyer, *Ole Rømers Adversaria* (Copenhagen, 1910).

115. In his lecture on Monday 13 December 1751, *loc. cit.* 46, item 15_{40}, Mayer remarks:

'—findet man durch Hülfe der Höhern Geometrie daß der Diameter des Aequatoris unserer Erde un dem 220 igsten Theil grösser ist, als die Axe derselben, welche von einem Pole der Erde bis zum andern gehet; Newton aber hat durch die Theorie welche sich auf die Umwälzung der Erde gründet, gefunden, daß der Diameter des Aequatoris der Erde zur Axe sich verhalten müsse wie 230 zu 229, also daß man auch hier keine grössere Übereinstimmung wünschen könnte.'

116. 'Dictata ad Geographiam (de figura telluris),' item 15_{21} in *loc. cit.* 46. This is an unfinished treatise, consisting of four chapters, composed on the basis of lectures given in the first half of 1760. Chapter I is the longest, and is entirely devoted to a mathematical demonstration of the shape of the Earth on the Newtonian System (i.e. as deduced from the inverse-square law of gravitational attraction): it concludes with a statement to the effect that if the Earth were fluid and its mass homogeneous, the ellipticity would be 1/230.

117. N. L. de la Caille, *Leçons élémentaire d'astronomie géometrique et physique* (Paris, 1746). This was translated into English by J. Robertson, under the title: *The Elements of Astronomy, deduced from Observations; and Demonstrated upon the Mathematical Principles of the Newtonian Philosophy: etc.* (London, 1750). The relevant sections in this translation are §§ 771–776, pp. 317–320. Three later French editions of La Caille's excellent survey of astronomical science appeared in 1755, 1761, and in 1779 with notes by De La Lande.

118. J. J. L. de Lalande, *Astronomie,* ii (Paris, 1771), pp. 494–498. See also the 2nd and 3rd editions of the same work.

119. In Mayer's unpublished treatise *Astronomie* which, being an improved version of his lecture notes composed sometime in 1754, is also preserved in *loc. cit.* 46, item 15_{40}, he defines the right ascension of a star as the angle between the hour-circle through the Vernal Equinox and that through the star, without specifying the direction in which it was to be measured. It would therefore be quite consistent with this definition to regard μ as the shorter of the two arcs measured along the celestial equator from ♈ to Q (the culminating point), as Mayer also appears to be doing in deriving the formula on p. 74 for the increase in the Moon's equatorial horizontal semidiameter; in which case the minus sign could be omitted. This is not so, however, if right ascension is defined (in accordance with convention) as increasing in *an easterly direction* from ♈ along the celestial equator.

120. In Mayer's notation, $\sin A^2 \equiv \sin^2 A$.

121. J. de Lalande, *Astronomie,* ii (3rd edition, Paris, 1792), p. 387.

122. J. P. Stern, *Lichtenberg: a Doctrine of Scattered Occasions reconstructed from his Aphorisms and Reflections* (Indiana, 1959), p. 79.

123. *Ibid.*, p. 80.

124. See L. C. Lichtenberg and F. Kries (eds.), *Lichtenbergs physikalische und mathematische Schriften,* iv (Göttingen, 1806), pp. 145–146, where Lichtenberg says: "Alle mathematischen Gesetze, die wir in der Natur finden, sind mir trotz ihrer Schönheit immer verdächtig. Sie freuen mich nicht. Sie sind bloss Hülfsmittel. In der Nähe ist Alles nicht wahr.'

125. 'Mathematische Untersuchung der Farben-Mischungen,' *loc. cit.* 46, item 15_{12}. See, in particular, §§ 3 and 4.

126. *Ibid.*

127. *Loc. cit.* 46, item 15_{13}. This text has been published by Forbes (*op. cit.* 44).

128. *Loc. cit.* 46, item 13. The contents of this manuscript have been discussed by Forbes (*op. cit.* 44).

129. J. H. Lambert, *Beschreibung einer mit dem Calauschen Wachse ausgemalten Farbenpyramide*... (Berlin, 1772) had developed the theory of colour-mixing independently and taken acount of the masses and strengths of colouring of the pigments in his treatment.

130. This notation is explained in the text of his Latin treatise (cf. IV).

131. I. Newton, *Opticks* (London, 1704).

132. *Goethe's Naturwissenschaftliche Schriften, II. Abtheilung,* 2. *Band zur Farbenlehre* (Weimar, 1890); an English translation by C. L. Eastlake entitled *Goethe's Theory of Colours,* was published in London in 1840.

133. W. Heisenberg, *Wandlungen in den Grundlagen der Naturwissenschaft,* 5. Auflage (Leipzig, 1944). Cf. chapter entitled: 'Die Goethesche und die Newtonsche Farbenlehre im Lichte der modernen Physik', (pp. 58–76).

134. *Loc. cit.* 46, item 15_{40}. Lecture dated Tuesday 18 January 1752.

135. 'Mayersches Zodiacal-Sternverzeichniss, auf den Anfang des Jahres 1800 reducirt, und durch Hinzusetzung der Längen und Breiten der Sterne, wie auch der jährlichen Veränderung in der geraden Aufsteigung und Abweichung, zum Gebrauch bequemer eingerichtet von J. A. Koch, M.D.', *Astronomisches Jahrbuch für das Jahr 1790 . . .* (Berlin, 1787), pp. 113–144.

136. *Astronomical Observations, made at Gottingen, from 1756 to 1761. By Tobias Mayer, Professor of Astronomy, and Member of the Royal Society of Sciences of Gottingen* (London, 1826). The Preface—or Advertisement, as it is called—contains reprints from the *Opera inedita* of treatise II 'Observationes Astronomicae Quadrante Murali Habitae in Observatorio Gottingensi', (pp. v–viii), and Lichtenberg's description of Mayer's thermometer (pp. ix, x) in 'Ad Commentationem I'.

137. Francis Baily, 'Mayer's Catalogue of Stars, corrected and enlarged; together with a Comparison of the Places of the greater part of them, with those given by Bradley; and a reference to every observation of every Star,' *Memoirs of the Astronomical Society of London* iv (1831), pp. 391–445.

138. *Tobias Mayer's Sternverzeichniß, nach den Beobachtungen auf der Göttinger Sternwarte in den Jahren 1756 bis 1760, neu bearbeitet von Arthur Auwers* (Leipzig, 1894).

139. E. Halley, 'Considerations on the Change of the Latitudes of some of the principal fixt Stars', *Phil. Trans. R. Soc.*, xxx (1718), pp. 736–738.

140. W. Herschel, 'On the proper Motion of the Sun and Solar System; with an Account of several Changes that have Happened among the Fixed Stars since the Time of Mr Flamstead', *ibid.*, lxxiii (1783), pp. 274–283.

141. J. L. de Lalande, *Astronomie*, iv (2nd edition, Paris, 1781), p. 685. The proper motions of two of the twelve stars (β and ε Cygni) listed in the *Opera inedita* refer only to a 44-year interval. The corresponding values given by de Lalande are linear extrapolations of these figures for a 50-year interval, so as to make them directly comparable with those for the other ten stars.

142. J. L. de Lalande, 'Mémoire sur les taches du soleil, et sur sa rotation', *loc. cit.* 47, *Année MDCCLXXVI* (Paris, 1779), pp. 457–514. The displacement of the solar system through space is discussed on pp. 513–514.

143. P. Prévost, 'Mémoire sur le mouvement progressif du centre de gravité de tout le Système solaire', *Nouveaux Mémoires de L'Académie Royale des Sciences et Belles-Lettres. Année MDCCLXXXI* (Berlin, 1783), pp. 418–421.

144. P. Prévost, 'Mémoire sur l'origine des vitesses projectiles, contenant quelques recherches sur le mouvement du Système solaire,' *ibid.*, pp. 422–462. See also Prévost's supplement to this article on p. 472.

145. See F. Baily, 'On The Proper Motion of the Fixed Stars,' *loc. cit.* 137, v (1883), pp. 147–170. On page 147 of this review article, Baily gives Mayer the credit for having been the first to draw astronomers' attention to the subject.

146. P. Prévost and F. Maurice, 'Über die eigene Bewegung einiger Sterne, zwischen 1756 und 1797,' *Astronomisches Jahrbuch für das Jahr 1805 . . .* (Berlin, 1802).

147. Lichtenberg to Dieterich; Kew, 30 October 1774. *Op. cit.* 93, Letter III, p. 200. The quote is taken from the translation of this letter by Mare and Quarrell, *op. cit.*, 77, p. 60.

148. Lichtenberg to [Christian Gottlob] Heyne; London, 16 March 1775. *Ibid.*, p. 85.

149. Lichtenberg to Schernhagen; Kew, 16 October 1775. *Op. cit.* 93, Letter 121, Vol. I, p. 234.

150. Lichtenberg to Dieterich; London, 13 October 1775. *Op. cit.* 93, Letter 120, p. 232.

151. See, for example, K. Przibram, 'Lichtenberg als Physiker (150 Jahre Lichtenbergsche Figuren)', *Naturwissenschaften*, xv (1927), pp. 423–425. A detailed account of these electrical figures in the light of modern atomic physics is given by the same author in H. Geiger and K. Scheel, *Handbuch der Physik*, XIV (Berlin, 1928).

152. Johann Christian Polycarp Erxleben (1744–1777), Ordinary Professor of Physics at Göttingen since 1775, wrote works on chemistry and natural history and a much read textbook of physics entitled *Anfangsgründe der Naturlehre* (Göttingen, 1772).

153. Immediately after Erxleben's death, his five weekly lectures were taken over by Kästner. By the summer of 1778, Lichtenberg was giving two of these on selected topics,

and by the winter semester 1781–1782 he had assumed complete control of the course. See Hahn, *op. cit.*, 80, p. 44.
154. G. Gamauf, *Lichtenbergs Vorlesungen über Erxlebens Anfangsgründe der Naturlehre* (Wien-Triest, 1808).
155. W. Grenzmann, *G. Chr. Lichtenbergs Gesammelte Werke*, 2 vols. (Frankfurt a/M, 1949).
156. Lichtenberg wrote: 'Tobias Mayer hatte hinten in eines seiner Bücher geschrieben: *quaeritur*, ist es besser, wenig und das deutlich zu wissen, oder viel und undeutlich?' (From what we know of Mayer it is evident that this question is meant to be a rhetorical one.) He then discusses Mayer's query and, not surprisingly, plumps for the former alternative; viz. that it is preferable to know a little, thoroughly. *Op. cit.* 49, pp. 50–51.
157. A full bibliography of G. C. Lichtenberg's works is given by: A. Schneider, *G. C. Lichtenberg* (Nancy, 1954), pp. 303–320.

Preface

158. In this context, the term 'Royal Society' does not refer to that of London—an error made by J. P. Stern (*op. cit.* 122, p. 20)—but is an abbreviation for the Royal Society of Sciences in Göttingen, founded in 1751. The first four volumes of the *Commentarii Societatis Regiae Scientiarum Gottingensis* were published in successive years between 1752 and 1755, but a dispute with the printer (cf. ref. 160) and the advent of the Seven Years' War caused a suspension in the publication of these translations until 1769/70 when Volume I of the *Novi Commentarii . . .* first appeared. Reports on the work of the Society during the intervening 14 years, and subsequently, are contained in the *Göttingische Anzeigen von gelehrten Sachen* (alias the *Göttingen gelehrte Anzeigen*).
159. Presumably Lichtenberg here has in mind the delays incurred through his measurements in Stade, and perhaps also the difficulties involved in editing Mayer's treatise on colour.
160. This arose over the bookseller's refusal to accept the small compensation offered by the Göttingen Society as a result of Mayer's retraction of the manuscript of *Methodus longitudinum promota* (*op. cit.* 35, pp. 1–31) so that it could be sent to the British Admiralty in support of his claim for a major Parliamentary award.
161. Introduction, p. 45. Besides the treatises cited in refs. 162, 163, 164, Poggendorff (*op. cit.* 98) lists the following three works as being destined for a second volume of Mayer's *Opera inedita*; yet they are no longer to be found among the manuscripts preserved in Göttingen:

De refractionibus astronomicis (1755).
De transmutatione figurarum rectilinearum in triangula (1755).
Artis qua picturae datae ectypa multiplicantur specimen exhibitum (1759).

It would appear to be more than a mere coincidence that the first scientific work of Mayer's eldest son (published in Altdorf and Nuremberg 1781) happens to bear exactly the same title as the first of these missing treatises! What would appear—judging from the abstract in *loc. cit.* 39 (1755), p. 29—to be a German version of the second, is still preserved among Mayer's papers under the title 'Über die Ausmessung der Vielecke durch Diagonalen' (cf. 15_{22}, *op. cit.* 46). Reference has already been made to the third in ref. 45.
162. *Op. cit.* 46. A summary of Mayer's views on terrestrial magnetism, based upon the abstracts cited in ref. 41, is given by: J. S. T. Gehler, *Physikalisches Wörterbuch*, vi (2nd edition, Leipzig, 1836), in Ch. xvii, pp. 1023–1147. This same chapter also contains an account of Euler's theory on which Mayer's was based.
163. The full title of this treatise, which Mayer read at a meeting of the Göttingen Scientific Society on 10 April 1756, is: 'Theoria motus Martis ex principio attractionis Newtonianae deducta' (cf. 15_{31} in *op. cit.* 46). See also *loc. cit.* 39 (1757), pp. 425–426.
164. Mayer 15_{23}, *op. cit.* 46 (cf. *loc. cit.* 39).
165. At the end of this roughly written draft (*ibid.*) there are diagrams of the astrolabe and a note from Kästner saying that the instrument is in his possession. See Plate XIV in the article by Forbes (*op. cit.* 44).
166. Lichtenberg must be alluding here to his remarks on Mayer's treatise on colour-mixing (cf. pp. 81–91).

167. J. A. Koch (*loc. cit.* 135) was the first to make the improvements to which Lichtenberg is referring (cf. Introduction, p. 43).
168. These are the errors cited for pp. 51, 52, 62, 66, in the *Corrigenda* on the last page of the *Opera inedita* (p. 110).
169. Christian Gottlob Heyne (1729–1812), an eminent colleague of Lichtenberg's in Göttingen. See ref. 108.
170. *Op. cit.* 19.
171. Perhaps Kästner had previously told Lichtenberg that he intended to have the catalogue of Mayer's writings reprinted in Part 2 of his collection of astronomical treatises which appeared in 1774 shortly after Lichtenberg had written this preface (cf. *op. cit.* 302). He may, however, by this time have decided against this, thinking that Lichtenberg himself would be able to include the list in a later volume of the *Opera inedita*. At any rate, whether there is any truth in these suppositions or not, the fact is that it is not included either in this or in any other of Kästner's later works.
172. The copper-engraving of Mayer's self-portrait—the only one in existence, according to F. X. von Zach, *Allgemeine Geographische Ephemeriden*, iii (1799), p. 116—was made by the same artist, Joel Paul Kaltenhofer (d. 1777), whom Lichtenberg had engaged to engrave the unfinished segments of Mayer's smaller lunar map (cf. Introduction, p. 18).
173. The preface was not, in fact, written in Göttingen, but en route to England (cf. Introduction, p. 20).

Treatise I

174. T. Brahe, *Astronomiae instauratae progymnasmata*, 2 vols. (Prague, 1602–1603).
175. Johannes Kepler (1571–1630), the German mathematician and astronomer who (after years of laborious calculation) discovered the three laws of planetary motion which bear his name. Newton was the first to demonstrate that all three could be deduced from the single hypothesis of a gravitational force between the Sun and a planet (or between a planet and its satellites) decreasing in proportion to the inverse-square of their distance apart.
176. This formula is an empirical one, which must have been based upon the assumption —presumably justified by observations made prior to 1755—that the temperature at latitude $\phi=\pm 45°$ is the arithmetic mean of those at the equator $\phi=0°$ $(=m)$ and either of the poles $\phi=\pm 90°$ $(=m-n)$; for $\{m+(m-n)\}/2=m-n/2=m-n(\frac{1}{2})^2=m-n\sin^2\phi$ where $\phi=\pm 45°$. Later observations in the northern hemisphere have shown that this formula is reliable only between latitudes 40° and 60° N.; outside this range, in more northerly latitudes, it predicts values which are much too high. See *Encyclopaedia Britannica*, xvii (7th edition, Edinburgh, 1842), article 'Temperature', p. 543.
177. P. Bouguer, *La Figure de la Terre . . .* (Paris, 1749), p. xxi, remarks: 'cependant le Thermomètre de M. de Réaumur n'y montait l'après-midi qu'à 26, 27 ou 28 degrés; le matin un peu avant le lever du Soleil, il se trouvait ordinairement a 19, 20 ou 21.' Thus 24°R. would seem to be a good approximation to the mean daily temperature on coastal planes near the equator in Peru, where Bouguer's measurements were made. Alexander von Humboldt, however, estimated this mean value to be 2°R. too high. (See *Encyclopaedia Britannica, ibid.*)
178. To the nearest degree, 49° N. is the latitude of Paris and 52° N. that of Göttingen.
179. By definition, the interval between the freezing and boiling points of water on the Réaumur and Fahrenheit temperature scales are 80°R. and $212°-32°=180°$F. respectively. Hence $1°\text{R.}\equiv 180/80=2\frac{1}{4}°\text{F.}$ and $24°\text{R.}\equiv 24\times 2\frac{1}{4}=54°\text{F.}$ However, as Lichtenberg points out in his commentary, Mayer's Réaumur scale differed from the conventional one in having $82\frac{1}{2}°$ between the two fixed points (cf. p. 118), thus yielding $52\frac{4}{11}°$ as the equivalent Fahrenheit temperature difference. The actual reading on the Fahrenheit scale corresponding to 24°R. on Mayer's scale is therefore $52\frac{4}{11}°+32°=84\frac{4}{11}°$, or 'around 84', as stated in the text.
180. The French mathematicians were Pierre Bouguer (1698–1758) and Charles Marie de La Condamine (1701–1774), the leading members of the French expedition to Peru in 1735 to measure the length of a degree of the meridian at the equator.

181. Bouguer, *op. cit.* 177, p. 1, states: '—l'intervalle . . . vertical entre les deux termes, le supérieur & l'inférieur, de la neige, est au moins de 11 ou 1200 toises dans la Zone torride; & qu'il y eut des montagnes assez hautes, on leur verroit une ceinture ou Zone de glace qui commenceroit à 2440 toises au dessus de niveau de la Mer . . .'.

182. Mayer's assumption of a mean temperature decrease of 1 °R. for an increase in altitude of 100 toises follows from his adoption of the result obtained by Bouguer and La Condamine in Peru (cf. ref. 180)—implicit in the statements quoted in refs. 177 and 181—that there is a drop in mean temperature from 24 °R. to 0 °R. from low-lying places to an altitude of about 2400 toises, coupled with his working hypothesis that the temperature decreases linearly with height.

183. This result is slightly erroneous, on account of an arithmetical error incurred by Mayer in interpolating from his table [Table 1]. The mean temperature corresponding to a latitude of $51\frac{1}{2}°$ should be $9\frac{2}{5}$ °R., not $9\frac{3}{5}$ °R. as stated; hence, if one accepts Mayer's assumption regarding the altitude correction, the mean temperature for Göttingen should be $8\frac{4}{5}$ °R.

184. The fact that both these dates are about 30 days from the summer and winter solstices respectively, indicates that for Göttingen the maximum and minimum temperatures occur when the Sun is $30 \times 360°/365\frac{1}{4} = 29{\cdot}6°$ (or approximately 30°) from these solstices. Since it is clear from the figures tabulated in Table 3 that this is indeed Mayer's meaning, the words 'trigesimum gradum' in the Latin text have been interpreted as 'thirty degrees' in the English translation.

185. Mayer appears to have been assuming that 'the Sun's degrees after a solstice' ($d_\odot$ say) is proportional to $\sin^2 \phi$. Since 51° 32′ was the latitude which he had found for Göttingen (cf. 37, pp. 446–448)

$$\frac{d_\odot \text{ at } \phi}{\sin^2 \phi} = \frac{30°}{\sin^2 51° 32'} = 49° \quad \text{or in general} \quad d_\odot = 49° \sin^2 \phi$$

The figures in Table 3 are, however, closely approximated by adopting 56° instead of 49° as the constant of proportionality, which would be consistent with $\phi = 47°$.

186. Expressed mathematically, this statement is equivalent to the expression $T = D \sin(\lambda_\odot - d_\odot) + T_0$, where T is the temperature at any time of year, when the Sun's celestial longitude is $\lambda_\odot$; and T_0 the mean temperature at an angular distance $d_\odot$ from either solstice.

187. The hypothesis must have been that the maximum annual temperature variation is directly proportional to the sine of latitude; for this gives:

$$\frac{\text{max. temp. variation at lat. } \phi}{\sin \phi} = \frac{\text{max. temp. variation at Göttingen}}{\sin 51° 32'}$$

or max. temp. variation at

$$\text{lat. } \phi = \frac{10°}{0{\cdot}783} \sin \phi = 12{\cdot}8° \sin \phi$$

on the Réaumur temperature scale. It is easily verified that the figures tabulated in Table 4 are in accordance with this formula.

188. In order to obtain the results tabulated in Table 5, Mayer has assumed the annual temperature variation to be sinusoidal.

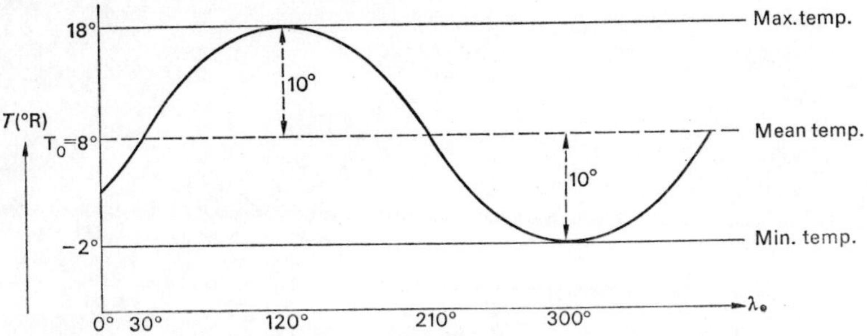

Fig. 20

In this particular case, for Göttingen, the mean and extreme temperatures are attained at times when the sun's longitude ($\lambda_{\odot}$) is 30°, 120°, 210°, 300°, as indicated in the accompanying diagram. The functional relationship between T and $\lambda_{\odot}$ is given by:

$$T = 10° \sin(\lambda_{\odot} - 30°) + 8° \quad \text{(cf. ref. 186).}$$

189. An inspection of the figures in Table 6, which again apply to Göttingen, reveals that the mean daily temperatures were found to occur near 9 a.m. and 9 p.m. only around the times of the equinoxes, being nearer 8 a.m. and 11 p.m. at summer solstice and near 11 a.m. and 7 p.m. at winter solstice.

190. The main points to note are that Mayer assumed *the same* maximum diurnal temperature variation of ±4°R. for all times of the year, with *fixed* maxima and minima at 2 p.m. and 4 a.m. respectively (cf. Table 6). Since the intervals between these times are 14 hours and 24 − 14 = 10 hours, the diurnal temperature variation cannot be expressed analytically by a simple sine curve.

Treatise II

191. John Flamsteed (1646–1719) was the first English Astronomer Royal and author of *Historia coelestis Britannica*, 3 volumes (London, 1725). The first two volumes contain the whole of Flamsteed's astronomical observations at Derby and Greenwich; while the third contains the *British Catalogue* of nearly 3000 star positions. The main authority on Flamsteed is Francis Baily, *An Account of the Rev. J. Flamsteed, the first Astronomer Royal . . . to which is added his British catalogue of stars, corrected and enlarged* (London, 1835).

192. Cf. ref. 36, which contains a description of this quadrant which had been approved by James Bradley and the 2nd Earl of Macclesfield. A diagram, which ought to have accompanied the description, is unfortunately missing.

193. J. Bird, *The Method of Constructing Mural Quadrants . . .* (London, 1768) contains more detailed information about the design and construction of the 8 ft. radius mural quadrant in the Royal Greenwich Observatory, of which that made for the Göttingen Observatory was an exact copy.

194. These observations constitute the basis of Mayer's catalogue of zodiacal stars (cf. IV).

195. See ref. 202.

196. Mayer again claims priority for the discovery of this error (which he goes on to explain in the text) in an unpublished letter to De la Caille, begun on 3 September 1758 (*loc. cit.* 27, p. 41).

197. Using the figures quoted in the text,

$$\text{Re} = \frac{\text{eE}}{\text{eF}} \cdot \text{ce} = \frac{30 \times 60 \times 60}{11} \times \frac{5}{4} \times \frac{1}{12^2}\ \text{ft} = 1\frac{1}{44} \times 1000\ \text{ft.} = 1023\ \text{ft}$$

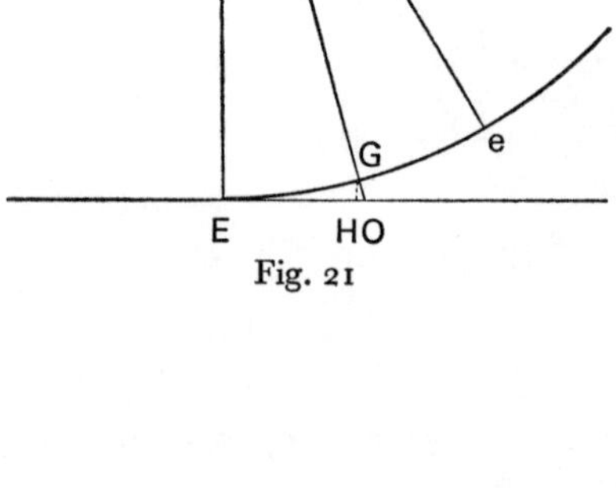

Fig. 21

(correct to the nearest foot). If the curvature is assumed to be circular (i.e. if Re is constant) within the 30° interval between E and e, the mid-point G of the arc Ee is such that $\widehat{ERG}=\widehat{GRe}=15°$. Produce RG to meet the tangent plane through E at O. Draw GH || RE to meet EO at H. HO is equal to the distance measured along the limb of the 6 ft radius quadrant corresponding to an error

$$eF=\frac{\frac{1}{2}eE.ce}{Re}=\tfrac{1}{2}\times 30\times 60\times 60\times\frac{1}{10\,000}=5{\cdot}4''$$

in the value actually obtained for a star of zenith distance EG due to the telescope being situated $1\frac{1}{4}$ inches from the limb of the quadrant. That is,

$$HO=6\times\frac{2\pi}{360}\times\frac{5{\cdot}4}{60\times 60}\text{ ft}=0{\cdot}157\times\frac{1}{1000}\text{ ft}$$

since the limb of the quadrant is circular.

$$GH=HO\cot\widehat{HGO}=HO\cot\widehat{ERG}=\frac{0{\cdot}157}{1000}\times\cot 15°\text{ ft}=\frac{0{\cdot}6}{1000}\text{ ft.}$$

This confirms Mayer's remark that the limb of his quadrant does not differ from a plane by even 1/1000 ft at the point mid-way between E and e if the curvature is assumed to be circular at this part of the scale.

198. That is, altitudes of different celestial bodies, referring to the same instant of time—as opposed to equal altitudes of the same body at different instants of time (cf. ref. 229).

199. From Mayer's text and the analysis in ref. 197, it should be evident that he is here referring to the variable local curvature *superimposed upon* the constant curvature of the 6 ft. radius limb, and not to the latter itself.

200. Namely, the rotational motion of the Earth on its axis. It is now known, however, that even this motion is subject to small irregular variations (resulting from changes in the Earth's internal constitution coupled with gravitational perturbations) whose amounts can only be empirically estimated with reference to another independent standard, viz., the assumed constant-frequency vibrations of a quartz-crystal clock.

201. The procedure of summing the three individual errors to obtain the total error is valid on account of the fact that each is associated with a displacement in one of three mutually perpendicular planes—a simplification which enables Mayer to derive his formula very easily by means of projective geometry, in which he was highly skilled (cf. Introduction, p. 22).

202. With regard to (a), Lichtenberg adds the following note on p. 20 of the Latin text: 'Concerning these and the observations which the author mentioned on p. 15 above, they seem [to be] those which are cited in the appendix.'

Treatise III

203. De Lalande (e.g. *op. cit.* 121, Ch. X, pp. 327–447) gives a detailed account of the methods of calculating eclipses which were in use during the eighteenth century. He distinguishes four types: the method of projection, used by de La Hire (*op. cit.* 205) and Cassini; the nonagesimal method used by de La Caille (*op. cit.* 117); that involving parallactic angles and the parallax in altitude, which was his own personal preference; and an elaborate analytic method developed by Dionis du Sejour in 19 memoirs of the Paris Academy (*loc. cit.* 142) for most years between 1764 and 1783 (inclusive).

204. In this connection, it is relevant to quote de La Caille's remarks on the graphical construction of a solar eclipse, which Robertson (*loc. cit.* 117, p. 319) has translated as follows:

'In this construction it is supposed, that the sun is almost at an infinite distance from the earth and moon; that the right lines by which the moon's diameter is projected on the plane of the figure are parallel, that the track of *Paris* seen from the sun is an ellipsis, that the moon's diameter seen from the earth is constant, that its motion is rectilinear and uniform, &c. These suppositions do not exactly agree with what passes in the heavens. Therefore it is evident, that supposing the elements taken from the tables to be perfectly accurate, yet the phases of the eclipses could only be deduced nearly. And although the errors arising from these suppositions oftentimes compensate one another, nevertheless it may be reckoned that they introduce an uncertainty of two minutes in the times determined by graphical operations, though done with the utmost care. 'Tis therefore needless to make use of trigonometric calculation to know the times with more accuracy, unless it were to correct the foregoing suppositions, which would be a tedious and laborious work; . . .'

205. P. de La Hire, *Tabularum astronomicarum pars prior de motibus solis et lunae, etc.* (Paris, 1687); 2nd edition, 2 parts (Paris, 1729).

206. 'Sur le calcul des projections en général, et en particulier sur le Calcul des Projections propres aux Eclipses du Soleil & aux Occultations des Etoiles fixes par la Lune,' *loc. cit.* 47, Année 1744 (Paris, 1748), pp. 191–238.

207. S. Klingenstierna, 'Methodus nova eclipses solares computandi in breves regulas redacta,' *Acta S. R. Scient. Upsaliensis 1742* (Stockholm, 1748), pp. 107–128.

208. See Introduction, pp. 25–28, for the derivation of the formula which follows.

209. Note, however, that the same symbol (μ) is subsequently used to denote the right ascension of the culminating point ($=\Upsilon Q$ in Fig. 10) whereas here it is ambiguously defined as the moon's hour angle *east or west* of the observer's meridian.

210. That is, the Nonagesimal—or point on the ecliptic lying above the horizon whose celestial longitude is 90° greater than that of the point of intersection of these two planes (cf. Fig. 10, p. 30).

211. *Aegidii Strauchii Tabulae per universam mathesin, inter quas etiam astronomicae, sphaericae et theoricae* (Wittenberg, 1662). A later edition, edited by Leonhard Christoph Sturm was published in Amsterdam in 1700.

212. *Christiani Grunebergii Encyclopaedia mathematica, in quâ astrognosiae, sphaericae, theoricae et eclipsigraphiae elementa traduntur* (Frankfurt and Viadrum, 1688).

213. 'In Parallaxin Lunae eiusdemque a terra distantiam inquisitio,' *loc. cit.* 27, 168.

214. The second term in this expression should vanish, since at the time of a solar eclipse the Moon is on, or very close to, the ecliptic where its celestial latitude $a=0$ (cf. Introduction, p. 36).

215. *Op cit.* 28.

216. Mayer made 15 observations with an 8 ft. focus tube from the beginning of the eclipse at $9^h\ 14^m\ 44^s$ to the end at $11^h\ 38^m\ 34^s$. He cites these in the reference given, together with 12 values of the difference in declination between $10^h\ 18^m\ 4{\cdot}2^s$ and $10^h\ 35^m\ 50^s$.

Treatise IV

217. This is the Newtonian interpretation of the prismatic spectrum (cf. *op. cit.* 131); however, as Mayer remarks in *op. cit.* 125 (§4): 'viel mehr prismatische Farben, als Newton deren gezählet hat, können angegeben werden . . . Man hat vielleicht nur sieben dieser Farben gezählet, weil man in der Sprache nicht mehrere bequeme Namen dafür gefunden hat, und weil man der Meynung war, es müsse zwischen den Farben und den Tönen einer Octave einige Ähnlichkeit seyn.'

218. Apart from his experiments on colour-mixing, Mayer was convinced of the validity of his own hypothesis by a phenomenon of a different nature, which he describes in *op. cit.* 125 (§3) as follows: 'Will man sie die drei Hauptfarben so viel möglich ist, abgesondert und rein sehen, so betrachte man durch das Prisma einen gegen das helle Licht des Himmels aufgehängten schwarzen Stab oder etwas ähnliches. Man wird auf der einen Seite desselben einen schönen blauen Streifen, auf der andern aber einen rothen und nächst an diesem einen gelben Streifen wahrnehmen; jede dieser Farben wird von der andern fast vollkommen abgesondert, und die gemischten Farben, Grün, Purpur, u.d.g., die im Regenbogen und in dem prismatischen Strahl noch nicht vorkommen, werden hier nicht zu sehen seyn.'

219. Unfortunately, Mayer nowhere defines what he means by a 'perfect mixture', although a qualitative definition of a 'perfect colour' is given in §23 of this treatise.

220. Cf. §§23–28. As mentioned in the Introduction (p. 40), Mayer's unpublished manuscript 'Vom Coloris', written a year after this Latin treatise was read at a public meeting in Göttingen, deals with the problem of shading. By this time, he had come to regard 'perfect black' as the mixture $b^8r^2g^2$ (cf. §13 for an explanation of this notation), and 'the complete shadow-colour' as $b^4r^4g^4$, the equal mixture of the three primary colours. See the discussion by Forbes (*op. cit.* 44).

221. This recommendation is explicitly referred to by Lichtenberg in his commentary (cf. p. 128); and it was a colour-triangle constructed from dry pigments which he showed to the Göttingen Scientific Society on 6 August 1774 (cf. ref. 292).

222. 'g' was used since it is the initial letter of the German word 'gelb' (yellow).

223. The calculation which now follows in the Latin text is erroneous in several respects, and has therefore been omitted from this translation and replaced by the correct solution (in italics.) When summing up the distributives for the colours red (r) and yellow (g), Mayer wrongly grouped a factor $4x$ with the term in g instead of with that in r. The corresponding sum for blue (b) is correct, but when equated to 3 and rearranged it should yield the equation $3x+y-8z=0$ and not $3x-2v-8z=0$. This latter mistake, however, would appear to have been typographical, since Mayer's remark that 'the sum of the second and third equations produces the first' applies to the correct equation (above) and not to that cited in the text. A further error, which must be ascribed to Mayer himself, is that the 'solution' which he provides satisfies the second equation only, and not the system as a whole (as it should). A second typographical mistake—the printing of the symbol x instead of r—appears in the final result. Lichtenberg was apparently unaware of these inconsistencies in the Latin text.

224. This fact also occurred independently to Johann Heinrich Lambert (*op. cit.* 129), after he had read the description of Mayer's colour-triangle published in the *Göttingen Anzeigen* (*op. cit.* 40), where no explicit reference is made to the pyramidal nature of the number 364.

225. At this point, the geometrical analogy breaks down. Mayer's statement can only mean that black and white are situated at the opposite vertices of a double tetrahedron whose other three vertices define the common base, or the plane of the colour-triangle; and this implies a separation between black and white equal to $\sqrt{3}$ times the separation of either from any of the three vertices of the colour triangle, which is inconsistent with a 12-fold variation in degree of blackness. It would, in fact, be $12\sqrt{3}$. One must adopt the *alternative—not consequent*—algebraic definition introduced by Mayer in §28 that black is negative white if one is to adopt this 12-fold variation and its corollary that there exist the same number of dark colours as pale ones (cf. §29). Neither Mayer, Lambert (*op. cit.* 129), nor Lichtenberg seem to have appreciated this point.

226. This is equivalent to saying that the combination of two indeterminate quantities can produce one which is determinate.

Treatise V

227. The lecture in which Mayer first announced his intention of preparing the catalogue in question, was delivered to the Göttingen Scientific Society on 6 November 1756. *Loc. cit.* 39 (1756), 1257–1258.

228. For details regarding Mayer's mural quadrant, see II (pp. 62–70) and refs. 36, 192, and 193.

229. The equal altitudes method for determining the direction of the observer's meridian was a common one used both by astronomers and by navigators at sea. Normally, it involved observing the Sun's altitude with an astronomical quadrant, nautical sextant, or similar instrument several hours before midday and noting the time on a reliable timepiece (not necessarily a chronometer). Then, in the afternoon, the time was noted when the Sun had sunk back to the same altitude. The mean of the two times gave the time of local apparent noon, when the Sun was on the observer's meridian. Mayer had simply to compare this time with that of the Sun's transit across the plane of his mural quadrant, to determine the latter's inclination to the meridian plane.

230. Mayer is, of course, referring to the Seven Years' War in general, and to the occupation of Göttingen by French troops in particular. For details of the local hardships to which he and his colleagues were then directly exposed, see for example Saathoff (*op. cit.*

21, Ch. 4, pp. 44–63); also: S. C. Hollmann, *Geschichte der Universität Göttingen* (Göttingen, 1787); A. Schöne, *P.of. Hollmann's Chronik; daraus: Die Universität im 7j.Kriege* (*Leipzig*, 1887).

231. As explained in the Introduction (pp. 41–43), this procedure involves a knowledge of the latitude and the obliquity of the ecliptic.

232. Cf. ref. 191.

233. *Eustachii Manfredii Ephemerides motuum coelestium ab anno 1726 ad annum 1750* (Bologna, 1725). New ephemerides for 1741 to 1750 based on the meridian of Bologna, were also published by Manfredi in 1739.

234. Cf. ref. 229.

235. For the latitude of Göttingen, this corresponds to the part of the scale between 15° and 62°.

236. One compendium which Mayer was desperately trying to obtain in 1753 (cf. Kopelevich, *loc. cit.* 25, pp. 353 or 355), was: W. Gardiner, *Tables of logarithms for all numbers, from 1 to 102, 100 and for the lines and tangents to every ten seconds of each degree in the quadrant* . . . (London, 1742). The tables of Strauch (*op. cit.* 211) and Gruneberg (*op. cit.* 212) were but two among many other aids to calculation which Mayer may have had in mind. The first work in which the theory of precession and nutation was rigorously developed was J. D'Alembert, *Recherches sur la précession des équinoxes, et sur la nutation de l'axe de la terre, dans le système newtonien* (Paris, 1749). Formulae for finding the effect of stellar aberration in celestial longitude were given by A. Clairaut, 'De L'Aberration de la Lumière des Planètes, des Comètes et des Satellites,' in *loc. cit.* 47, *Année* 1746 (Paris, 1751), pp. 539–586. A comprehensive list of relevant books and articles can be found in Jerôme de La Lande, *Bibliographie Astronomique* . . . (Paris, 1803).

237. This revised figure was 24″ less than the result which Mayer had published five years earlier in *loc. cit.* 28, iii (1754), 448.

238. Mayer's Table of Refractions is published on p. xxxii of *op. cit.* 35. The original formula on which it is based, which was 'deduced from theory' (cf. *ibid.*, p. 118), is quoted on *ibid.*, p. 117. A comparison of the refractions calculated by Mayer, de La Caille and Bradley is given by R. Wolf, *Handbuch der Astronomie, ihrer Geschichte und Litteratur*, 2 vols. (Zürich, 1890), p. 270. It shows that the results of Mayer and Bradley were superior to those of de La Caille and very close approximations to the truth up to a zenith distance of 80°. Bradley's formula then gave a somewhat better representation of the observed refraction in the range from 80° to 90°.

239. In his solar and lunar tables (*op. cit.* 35), Mayer used a value of 50·3″ for the annual precession of the equinoxes when computing the mean motions of the Sun and Moon, and remarks (*ibid.*, p. 107): 'The maximum of nutation 9″·6 . . . exceeds a little that which is commonly assumed, the laws of attraction requiring such increase, and the observations at the same time favouring it.' Perhaps he calculated the values of stellar aberration with the aid of the equations given by Clairaut (*op. cit.* 236), or from those of Durdus (cf. de Lalande, *op. cit.* 236, p. 442).

240. This would appear to be a guess. In fact, if the observational errors are assumed to be randomly distributed about the mean value, the uncertainty in a star position deduced from ten observations should be over 3″, and not 2″ as Mayer claims.

241. Lichtenberg, in a letter to Johann Friedrich Wurm dated Göttingen, 23 December 1787 (*loc. cit.* 93, Letter 522, Vol. II, p. 315) remarks that he had discovered that the stellar magnitudes quoted by Mayer in his star catalogue had been taken from Flamsteed's catalogue (*op. cit.* 191).

242. This was done by J. A. Koch, *op. cit.* 135.

243. Mayer would appear to have forgotten that the obliquity of the ecliptic was also implicit in the values of the right ascension of the Sun—hence of the stars—which he had derived (cf. Introduction, p. 43).

Treatise VI

244. The lecture in which Mayer presented his new star catalogue (cf. V) to the Royal Society of Sciences in Göttingen was delivered on 7 April 1759.

245. Edmond Halley (1657–1742) succeeded Flamsteed as Astronomer Royal at Greenwich on the latter's death in 1720. A popular account of his life and work is to be found in: Armitage, *Edmond Halley* (London and Edinburgh, 1966).

246. The most reliable text of Ptolemy's *Almagest*, the work to which Mayer is here referring, is: *Claudii Ptolemaei opera quae exstant omnia*, Vol. I *Syntaxis mathematica*, edidit J. L. Heiberg (Leipzig, 1898).

247. See Halley, *op. cit.* 139. The relevant passage is also quoted by A. Armitage, *op. cit.* 245.

248. Jacques Cassini (1667–1756), son of the famous Giovanni Domenico Cassini, observed that the latitude of Arcturus had changed by 5′ in 150 years, while the nearby η Bootis had not. See K. G. Jones, 'The Search for the Nebulae—IV,' in *J. Br. astr. Ass.*, lxxix (1968), 23.

249. See *Recueils d'observations faites en plusieurs voyages, par ordre de S.M., pour perfectionner l'astronomie et la géographie; avec divers Traités astronomiques, par MM. de l'Académie royale des sciences:* 6. 'Observations astronomiques faites aux costes septentionales de France, pendant l'année 1681, par MM. Picard et De La Hire' (Paris, 1693).

250. P. Le Monnier, *Histoire céleste, ou recueil de toutes les observations astronomiques faites par ordre du roy* (Paris, 1741).

251. Ole Roemer (1644–1710) was the Danish astronomer who is remembered principally as the discoverer of the finite velocity of light, which was suggested to him by observations of the eclipses of Jupiter's satellites. Unfortunately, most of his astronomical observations perished in the great fire in Copenhagen on 21 October 1728, with the notable exception of those published in: J. G. Galle, *O. Roemeri triduum observationum astronomicarum a. 1706 institutarum* (Berlin, 1845). Mayer knew of Roemer's *Triduum* . . . from Horrebow (*op. cit.* 252).

252. See P. Horrebowii, *Basis Astronomicae* . . . (Copenhagen, 1735) which is entirely devoted to a fundamental discussion of Ole Roemer's astronomical instruments, technique of observing, etc. The values of the right ascensions and declinations which Mayer employed in his investigation of proper motion, are listed under the title of 'Triduum Observationum Tusculanarum Roemeri' on pp. 167–208 in the third volume of the later (1741) edition of Horrebow's works (*op. cit.* 112).

253. Mayer is referring here to Roemer's 'Rota meridiana', which is described by G. van Biesbroeck in *op. cit.* 111, pp. 354–358; and, of course, by Horrebow (*op. cit.* 112 and 252).

254. *Astronomiae Fundamenta, Novissimus Solis et Stellarum Observationibus stabilita Lutetiae in Collegio Mazarinaeo et in Africa ad Caput Bonae-Spei peractis a Nicolao Ludovico De La Caille* (Paris, 1757). See Francis Baily, 'La Caille's Catalogue of 398 Principal Stars . . .', *loc. cit.* 137, v (1833), 93–124.

255. Literally: '—after ten lustrums have elapsed.' The lustrum was a sacrifice made by the Romans after every five years.

Commentary on I

256. See E. G. R. Taylor, *The Mathematical Practitioners of Hanoverian England*, 1714–1840 (Cambridge, 1966).

257. Abbé Jean Antoine Nollet (1700–1770) wrote a large number of *Mémoires* to the Paris Academy on physics generally and on his electrical experiments in particular. A text book of his, with which Lichtenberg must certainly have been familiar, is: *Leçons de Physique expérimentale*, 2 vols. (Paris, 1743).

258. The German translation of du Crest's writings on the thermometer and barometer is entitled: *Michael du Crest kleine Schriften von den Thermometern und Barometern* (Augsburg, 1765).

259. J. A. de Luc, *Recherches sur les modifications de l'Atmosphere* . . ., 2 vols. (Geneva, 1772).

260. G. Toaldo, *Della vera influenza degli astri delle stagioni e mutazioni di tempo, saggio meteorologico* . . . (Padua, 1770). This work was translated into German by I. G. Stendel (Berlin, 1777) and into French by J. Daquin (Chambéry, 1784). A new Italian edition, containing much additional data, was published by the author himself in Venice, 1786.

261. J. H. Lambert, 'Exposé de quelques Observations qu'on pourroit faire pour répondre du jour sur la Méteorologie', *Nouveaux Mémoires de l'Académie Royale des Sciences et Belles-Lettres. Année MDCCLXXI* (Berlin, 1773), pp. 60–65. Also, 'Observations sur l'Influence de la Lune dans le poids de l'Atmosphere', *ibid.*, pp. 66–73.

262. Le Pêre Cotte, *Traité de Météorologie* . . . (Paris, 1774).

263. 'Table de comparaison des Observations Météorologiques, faites en [62] différentes Villes', *ibid.*, pp. 386–387.

264. René Antoine Ferchault de Réaumur (1683–1757) described a thermometer in a pamphlet entitled 'Explication des principes établis par M. de Réaumur pour la construction des Thermomètres dont les degrés soient comparables' [Paris, 1725]: it was constructed on the principle that the freezing point of water was 0° and each degree was one-thousandth of the volume contained by the bulb and tube of his thermometer up to the zero mark. It so happened that the coefficient of expansion of the particular quality of alcohol which he employed, made the boiling point of water 80°. Thus 80° Réaumur are equivalent to 100° Centigrade or 1 °R. $\equiv 1\frac{1}{4}$ °C.

265. Cotte, *loc. cit.* 263, p. 386. The maximum and minimum thermometric and barometric readings for Mexico (latitude 20° 0′) are given as $25\frac{1}{2}$° and 8 °R., 21·5 and 21·1 inches respectively. These are the figures from which Lichtenberg derived this conclusion.

266. See *ibid.* The corresponding figures for Paris (latitude 48° 50′) are 30° and $-15\frac{1}{4}$ °R., 28·75 and 26·25 inches respectively.

267. *Ibid.*, p. 337. The title of the extract, pp. 336–342, is: 'Observations faites à Mexico [lat. 20° N., long. 106° W.] depuis le mois d'Avril 1769 jusqu'au mois de Décembre suivant, par Don Alzate y Ramirez, Chapelain du roi d'Espagne, Correspondant de l'Academie Royal des Sciences.'

268. *Ibid.*, pp. 365–366 (for Wilna). The title of this extract is: 'Observations faites à l'Observatoire royal de Wilna [lat. 54° 41′ N., long. 23° 7′ E.] en Pologne au collège des Jésuites, depuis le mois d'Avril 1770, jusqu'au mois de Mars 1771, par le P. Poczobut, Astronome du roi de Pologne.' Cotte remarks on p. 384 that the Petersburg results were collected from a variety of sources.

269. G. Gamauf, *Erinnerungen aus Lichtenbergs Vorlesungen über die Naturlehre*, Vol. 3 (Vienna and Triest, 1812), contains a section entitled 'Vom Thermometer', pp. 77–123, in which the thermometric scales of Fahrenheit, Réaumur, Deluc, Delisle and Celsius are respectively discussed. The description of the mean annual temperature scale is given by Lichtenberg in the present text. See also Cotte, *op. cit.* 262, pp. 99–142.

270. In fact, $1 : \sqrt{2} = 1000 : 1414$, which differs significantly (viz. by 3·4%) from the corresponding ratio of 1000 : 1380 adopted by Mayer.

271. R. Bonne (1727–1795), *Atlas maritime ou Cartes réduites de toutes les côtes de Frances* (Paris, 1762). See de Lalande, *op. cit.* 141, §2242, pp. 702–703.

272. Lichtenberg gives as his reference *Chym. Ed. Lips.* 1732, Vol. 1, p. 386.

273. Guillaume Amontons (1663–1705): 'Discours sur quelques Propriétés de l'Air, & le Moyen d'en connoître la Température dans tous les Climats de la Terre,' *loc. cit.* 47, *Année* MDCCII (Paris, 1743), pp. 155–173. Amontons describes his thermometer in *ibid. Année* MDCCIII (Paris, 1720), pp. 50–56. Its zero point was the (unattainable) temperature at which the 'spring' of the air was zero [about −240 °C.] The melting point of ice was at $51\frac{1}{2}$°, and the boiling point of water at 73°. If the zero were denoted alternatively by the number 1000, the boiling point would be $1000 \times 73/51\frac{1}{2} = 1418$, yielding for the expansion of air an accurate approximation to the ratio $1 : \sqrt{2}$, and an interval between these fixed points exactly $1\frac{1}{10}$ times that given by Mayer.

274. The foregoing remarks should be regarded merely as a justification for Mayer's adoption of his air-thermometer scale, and not as an indication of his manner of reasoning. As Lichtenberg himself points out, that scale was almost certainly based upon Mayer's own experiments.

275. De Luc, *op. cit.* 259, ii, §835 et seqq., pp. 263–264.

276. *Ibid.*, p. 264.

277. This ratio is given as 33 : 35 in the Latin text: it appears to have been a printing error.

278. De Luc, *op. cit.* 259, ii, p. 263.
279. *Ibid.*

Commentary on II

280. Mayer states in II that he turned the quadrant to face north on the 21 July [1756], and found that the zenith distances were 5″ greater than they ought to have been. The quadrant was restored to its former position on 6 August [1756], which marks the beginning of 'the epoch of the well-adjusted quadrant' (cf. p. 120).

Commentary on IV

281. J. H. Lambert, 'Mémoire sur la Partie Photométrique de l'Art du Peintre', *Histoire de L'Académie Royale des Sciences et Belles-Lettres. Année MDCCLXVIII* (Berlin, 1770), pp. 80–108.
282. Lichtenberg is here referring to colours obtained as the result of chemical action, which Mayer would appear to be excluding from consideration in §10 of IV (although the wording of that section is somewhat ambiguous as far as the effect of heat upon a substance is concerned).
283. Lichtenberg fails to recognize Mayer's distinction (in §27 of IV) that light cannot make a colour which is intrinsically dark, or mixed with black, any brighter.
284. This follows from the fact that the area of the colour triangle is one-half of the area of a square constructed on the same base, and that the length of that base increases in direct proportion to n.
285. Lambert (*op. cit.* 129, pp. 71, 72) generalises one of his conclusions by defining m parts by weight of dark red, n of dark Berlin blue, and p of amber as having the same strengths of colouring. If r, b, g are used to denote these three pigments respectively, colours in the basic triangle used by Lambert would be represented in accordance with Lichtenberg's considerations by the formula $A=\rho \mathrm{r}^m+\beta \mathrm{b}^n+\gamma \mathrm{g}^p$. The formula quoted by Lichtenberg in the text is of exactly the same form, except that $m+p$ and u have been used instead of n and p respectively (possibly because a different meaning has already been attached to n). The negative sign which appears in front of u on p. 95 of the original Latin text is a typographical error, and has therefore been omitted in the English version.
286. Lichtenberg probably also knew of Mayer's artistic talent through his contact with Kästner and with others who had known Mayer in Göttingen. Mayer's autobiographical account of his early childhood years, published by Benzenberg (*op. cit.* 11, pp. xxxv–li)—a manuscript copy is still preserved in the Göttingen University Library (*Cod. MS. Hist. lit.* 18)—contains a reference to it which Kästner introduced into his elegy (*op. cit.* 19). The contents of the *Mathematical Atlas* (cf. Introduction, p. 2), and the numerous maps which Mayer made for the Homann cartographic bureau (cf. ref. 20), also constituted ample evidence of his skill in drawing and design at subsequent phases throughout his youthful career.
287. This is composed of arsenic, sulphur and earth, and is a brilliant yellow colour.
288. The *Göttingen Anzeigen* . . . for 1758 (cf. *op. cit.* 40).
289. Literally: 'raised to higher orders'.
290. *Op. cit.* 129.
291. In a triangle like Mayer's for which $\rho+\beta+\gamma=6$, the primary colours are b^6, r^6, g^6 and one may write:

$$\mathrm{r}^6$$
$$\mathrm{r}^5\mathrm{b}^1 \quad \mathrm{r}^5\mathrm{g}^1$$
$$\mathrm{r}^4\mathrm{b}^2 \quad \mathrm{r}^4\mathrm{b}^1\mathrm{g}^1 \quad \mathrm{r}^4\mathrm{g}^2$$
$$\mathrm{r}^3\mathrm{b}^3 \quad \mathrm{r}^3\mathrm{b}^2\mathrm{g}^1 \quad \mathrm{r}^3\mathrm{b}^1\mathrm{g}^2 \quad \mathrm{r}^3\mathrm{g}^3$$
$$\mathrm{r}^2\mathrm{b}^4 \quad \mathrm{r}^2\mathrm{b}^3\mathrm{g}^1 \quad \mathrm{r}^2\mathrm{b}^2\mathrm{g}^2 \quad \mathrm{r}^2\mathrm{b}^1\mathrm{g}^3 \quad \mathrm{r}^2\mathrm{g}^4$$
$$\mathrm{r}^1\mathrm{b}^5 \quad \mathrm{r}^1\mathrm{b}^4\mathrm{g}^1 \quad \mathrm{r}^1\mathrm{b}^3\mathrm{g}^2 \quad \mathrm{r}^1\mathrm{b}^2\mathrm{g}^3 \quad \mathrm{r}^1\mathrm{b}^1\mathrm{g}^4 \quad \mathrm{r}^1\mathrm{g}^5$$
$$\mathrm{b}^6 \quad \mathrm{b}^5\mathrm{g}^1 \quad \mathrm{b}^4\mathrm{g}^2 \quad \mathrm{b}^3\mathrm{g}^3 \quad \mathrm{b}^2\mathrm{g}^4 \quad \mathrm{b}^1\mathrm{g}^5 \quad \mathrm{g}^6$$

Lichtenberg's triangle is found by weighting the distributives of b, r, g, in the above triangle in the ratio 3 : 6 : 1. For example, a colour-mixture such as $\mathrm{r}^1\mathrm{b}^3\mathrm{g}^2$ becomes

$r^{1\times3}b^{3\times6}g^{2\times1}=r^3b^{18}g^2$; and the three primary colours in Lichtenberg's triangle are $R=r^{6\times3}=r^{18}$, $B=b^{6\times6}=b^{36}$, $G=g^{6\times1}=g^6$, as is easily verified by inspecting Table 4.

292. *Loc. cit.* 39 (1774), pp. 905–907.

293. *Op. cit.* 129; see, in particular, pp. 108–126.

294. Translated from the French quoted by Lichtenberg from p. 107 of the Paris (1756) edition of Le Blon's work (cf. footnote on p. 132).

295. This is stated in *loc. cit.* 39 (1759), p. 402. By means of this art, coats of arms and oil paintings of all kinds and sizes could be reproduced as often as desired so that copies were exactly identical in colour, brightness, detail, lifetime, etc. See also ref. 44.

296. This work of art is referred to by Mayer's son in a letter to Lambert dated Göttingen, 2 November 1774 (*op. cit.* 45), where he states that it had originally been three inches thick; but so many French officers (billeted in Mayer's home during the Seven Years War) had cut specimens off it, that it was then reduced to only one inch in thickness. This circumstance would account for the state of the picture better than the reason which Lichtenberg is here proposing.

Remarks on The Selenographic Table

297. Lichtenberg was later to modify his opinion after seeing the copper engraving of Mayer's smaller lunar map made by Georg Heinrich Tischbein (1755–1848) for J. H. Schröter's *Selenotopographische Fragmente* . . . (Lilienthal, 1791). See his letter to Schröter, dated 21 April 1791, published in *op. cit.* 93, Vol. III, p. 234.

298. Cf. *op. cit.* 14.

299. The date 1760 given in the Latin text has been corrected to 1750. Cf. *op. cit.* 16.

300. 'Vorstellung der in der Nacht zwischen den 8 und 9 August 1748 vorfallenden partialen Mond Finsternis . . .' published as Tab. XXI in J. G. Doppelmaier, *Atlas Coelestis* (Nuremberg, 1752).

301. See Mayer, *op. cit.* 12.

302. A. G. Kästner, *Astronomische Abhandlungen zu weiterer Ausführung der astronomischen Anfangsgründe. Zweyte Sammlung* (Göttingen, 1774). Cf. pp. 272–278.

303. Lambert, *op. cit.* 12.

304. 'Bericht von den Mondskugeln . . .' (Nuremberg, 1750). See refs. 16 and 299.

305. A photolithographic reproduction of all these drawings was prepared in Dresden at the instigation of Wilhelm Klinkerfues, Director of the Göttingen Observatory, and published under the title: 'Tobias Mayer's grossere Mondkarte nebst Detailzeichnungen, zum ersten Male herausgegeben von der Königlichen Sternwarte zu Göttingen' (Göttingen, 1881). Figs. 2 and 3 in the present text appear, together with one other drawing of a small area of the Moon's disc near the craters Clausius and Regiomontanus, in: E. G. Forbes, 'Tobias Mayers Mondkarten', *Sterne und Weltraum*, viii (1969), 36–39.

306. Further details concerning this can be found in Herrmann, *op. cit.* 18, which also includes a photograph of Tischbein's copy (cf. ref. 297).

307. See ref. 63.

308. The incomplete state of the project at the time of Mayer's death is commented upon in *Erde- und Himmelsgloben, ihre Geschichte und Konstruction, nach dem Italienischen Matteo Fiorinis frei bearbeitet von Sigmund Günther* (Leipzig, 1895), pp. 128–131. It was left to the Englishman John Russell, in his book *A Description of the Selenographia: an apparatus for exhibiting the phenomena of the moon* . . . (London, 1797) to achieve what Mayer had intended. Russell used Mayer's principle of the micrometric determination of numerous small but well-defined craters, and indicated the unequal heights of individual lunar mountains by means of different kinds of shading.

309. Mayer's sensitivity on this point is understandable in the light of contemporary events both in Nuremberg and in Göttingen, details of which are to be found in the articles by Ruge (*op. cit.* 17) and Sandler (*op. cit.* 20). The unpublished documents and correspondence in the *Personalakte* of J. M. Franz and G. M. Lowitz (*loc. cit.* 43) at the University Archives, Göttingen, reveal that Mayer was more deeply involved in the cartographic projects of these two men—his two brothers-in-law—than the abovementioned printed sources tend to suggest.

310. 'Erklärung und Gebrauch der Mondcharte, durch Herrn Lambert,' *Astronomisches Jahrbuch für das Jahr 1776* (Berlin, 1774), pp. 134–154. In this article, written prior to the publication of the *Opera inedita*, Lambert remarks that he observed between 200 and 250 features on the Moon's surface and followed Mayer's method (cf. *op. cit.* 14) to determine their selenographic co-ordinates. He lists on pp. 151–152 his results (to the nearest degree) for 66 prominent lunar markings.

NAME INDEX

Alzate, D. 115, 116
Amontons, G. 117, 158
Anson, Lord 7, 11

Baily, F. 43, 44, 148, 152, 157
Bernoulli, J. 45, 144
Best, W. P. 6, 10, 11, 12
Biesbroeck, G. van 24, 146
Bird, J. 7, 17, 62, 152
Boerhaave, H. 117
Boie, H. C. 46
Bonne, R. 117, 158
Boston, Lord 15, 45
Bradley, J. 7, 10, 17, 143, 148, 152, 156
Brahe, T. 54, 150
Brander, G. F. 2, 137

Campbell, J. 13
Cassini, J. 109, 153, 157
Clairaut, A. vii, 5, 6, 13, 143, 156
Cook, J. 9
Cotte, P. 114, 115, 158
Coulomb, A. 8, 144

Dagoty, G. 133
D'Alembert, J. vii, 6, 156
Delambre, J. 21, 142
De Luc, J. A. 45, 114, 118, 119, 157, 158
Dieterich, J. C. 18–20, 25, 33, 45, 46, 146, 148
Du Crest, M. 114, 157

Einstein, A. 44
Erxleben, J. C. P. 46, 148
Euler, L. vii, 5, 6, 8, 13, 149

Flamsteed, J. 62, 63, 93, 109, 148, 152, 156, 157
Forster, G. 46
Franklin, B. 9
Franz, J. M. 5, 142, 160

Gamauf, G. 46, 149, 158
Gauss, K. F. 16, 145
Geiger, G. 1, 141
George III 5, 12, 13, 15, 16, 18, 45, 46
Goethe, J. W. von 41, 147
Grant, R. 14
Green, C. 11, 12
Gruneberg, C. 76, 154, 156
Günther, G. C. 134

Hahn, P. 19, 145, 146, 148
Halley, E. 43, 109, 148, 157
Harrison, J. 7, 11, 13
Heisenberg, W. 41, 147
Hell, M. 19, 45
Herschel, W. 44, 148
Hessen-Darmstadt, Duke of 15
Hevel, J. 2, 137
Hornsby, T. 45
Horrebow, P. 111, 146, 157

Kästner, A. G. 4, 10, 12, 16–18, 51, 52, 116, 137, 142, 148–150, 159, 160
Kattenhofer, J. P. 18, 136, 146, 150
Kandler, G. D. 1
Kepler, J. 54, 71, 150
Klingenstiern, S. 73, 154
Koch, J. A. 43, 148, 149, 156

Lacaille, N. L. de 30, 73, 111, 147, 152, 153, 156
Lagrange, J. L. de vii
La Hire, P. de 109, 153, 154, 157
Lalande, J. de 11, 19, 30, 38, 44, 45, 147, 148, 153, 156
Lambert, J. H. 17, 19, 40, 114, 122, 124, 127, 128, 130–133, 137, 139, 142, 144, 145, 147, 155, 158–161
Laplace, P. S. de vii
Le Blon, I. C. 131–134, 160
Leibniz, G. W. 9
Le Monnier, P. 109, 157
Lichtenberg, G. C. ix, 4, 9, 15–21, 24, 25, 33, 39, 40, 45–47, 142, 144–150, 153, 155–160

Manfredi, E. 93, 156
Mare, M. and Quarrell, W. H. 45, 145, 148
Maskelyne, N. 11–14, 17, 45, 143–145
Maupertuis, P. von 5, 143
Maurice, F. 44, 148
Mayer, J. T. (Tobias's son) 134, 142, 144, 145
Mayer, Maria V. 5, 10–13, 45, 142
Mayer, T. vii–ix, 1–22, 24, 25, 28–30, 33, 36–47, 51, 52, 113, 115–119, 121, 122, 124–130, 133–139, 141–161
Meister, G. 9
Michaelis, J. D. 6, 10–12, 143, 144
Munchhausen, G. A. von 12, 18, 142

Newton, I. vii, 8, 39, 41, 82, 147, 150, 154
Niebuhr, C. 45
Nollet, J. A. 114, 157

Pfeffel, J. A. 2
Picard, J. 109, 157
Poisson, D. vii
Prévost, P. 44, 148
Price, R. 45
Priestley, J. 45
Ptolemy, C. 109, 157

Riccioli, G. B. 2, 137, 140
Robison, J. 11
Roemer, O. 24, 111, 146, 157

Schernhagen, C. 17, 145, 146, 148
Schröter, J. H. 18, 146, 160
Schumacher, H. C. 16, 145
Segner, U. von 5
Sisson, J. 16
Stephens, P. 11, 12
Strauch, A. 76, 154, 156
Strömgren, E. 24, 146

Tischbein, G. H. 18, 160
Toaldo, G. 114, 157

Wolff, C. von 9

SUBJECT INDEX

Aberration 93, 95, 109, 114, 156
Admiralty 10, 11, 149
Altitude (above sea-level) 55, 57, 115, 116, 119, 121, 151
, of a heavenly body 65, 66, 74, 78, 92–94, 139, 153, 155
, of Nonagesimal 31, 34, 75, 76, 78
Atmosphere 53, 54, 57, 92, 95, 113, 115, 157, 158
Augmentation 25–28, 38

Barometer 53, 92, 114, 115, 119, 121, 157
Berlin 122, 143, 144, 147, 148, 158
, Academy of Sciences 5
, Academy *Mémoires* 122, 158, 159
Black 40, 81, 83, 90, 91, 123–125, 155, 159
Blue 40, 81–89, 122–131, 154, 155, 159
Board of Longitude, *see* Longitude, Commissioners (or Board) of

Chronometer 13, 144, 155
Cinnabar 40, 87, 125, 127, 128, 130
Colour, perfect 88–91, 124, 154
, primary 40, 81–87, 89, 123, 126–132, 154, 155, 160
, secondary (or composite) 81–88, 90, 122, 123, 126, 127, 129–131, 154, 155
Colour-mixing viii, 8, 19, 20, 39, 40, 81–90, 122–133, 143, 147, 149, 154, 159
-pigment 8, 40, 83, 86–88, 125–131, 133, 143, 147, 155, 159
-pyramid 40, 90, 122, 124, 127, 133, 147
-triangle 45, 52, 86, 122–133, 155, 159, 160
Co-ordinates (various) viii, 3, 4, 16, 18, 34, 43, 44, 96, 111, 140, 161
Cosmographical Society 136, 142
Culminating point 31, 76, 147

Declination of heavenly body 3, 7, 21, 26, 28, 41–44, 52, 69, 74, 75, 93, 95–108, 110–112, 154
, (dip) of magnetic needle 8

Earth 2, 3, 27, 28, 33, 44, 54, 58, 59, 68, 71–74, 95, 110, 112, 136, 153
Earthquake 16, 18, 19, 146
Eclipse, lunar vii, viii, 2, 16, 136, 139, 160
, of Jupiter's satellite(s) 11, 16, 157
, solar vii, viii, 2, 17, 34–39, 71–74, 77, 78, 80, 153, 154
Ecliptic 30, 31, 34, 41, 42, 58–60, 74, 78, 94, 96, 109, 154, 156

Elevation of pole, *see* Latitude, terrestrial
England 6, 11, 17, 19, 132, 150
Equator, celestial 28, 30, 31, 41, 69, 74, 76, 109
, lunar 3, 136, 139
, terrestrial 54, 60, 150
Equinox 30, 31, 42, 59, 63, 93, 147, 152
Esslingen 1, 2, 141

France 17, 132

Georg-August Academy, *see* Göttingen University
Göttingen viii, 5, 7, 8, 10, 12, 15–18, 46, 57, 76, 78, 80, 130, 133, 141–146, 148, 150–152, 155, 156, 159, 160
, *Anzeigen* 18, 149, 155
, *Commentarii* (transactions) 5, 6, 17, 18, 51, 76, 78, 80, 143, 149
, Observatory 5, 7, 10, 12, 17, 21, 45, 51, 62, 63, 93, 152
Scientific Society ix, 10, 17, 51, 92, 130, 134, 148, 149, 155, 156
, University 1, 4, 5, 15, 16, 133, 142–145, 159

Hadley's (sea-)octant, *see* Quadrant, nautical
Hanover 5, 12, 16, 17, 20, 142–145
Hanoverian Government 6, 10–12, 17, 19, 51
Harrison's timekeeper (or watch) 7, 11, 13
Homann Cartographic Bureau 2, 5, 8, 142, 159
Horizon 7, 27, 30, 31, 36, 72, 74, 78, 154

Instrument 7, 16, 62–64, 67, 69, 70, 92, 93, 111, 113, 114, 143, 145
Instrumental error viii, 7, 17, 21, 24, 44, 64–70, 94, 95, 113, 114, 153

Latitude, celestial 6, 25, 34, 37, 38, 52, 74, 76–80, 96, 154, 157
, selenographic 3, 136, 139, 140
, terrestrial (or geographic) 3, 8, 11, 16, 25–29, 31, 33, 42, 54–57, 59, 72, 75–78, 95, 118, 151, 156
Libration 3, 12, 137–139, 142
London 6, 15–17, 19, 132, 143, 144, 147, 148, 152, 156
Longitude, at sea 6, 7, 14, 143, 145
, celestial viii, 6, 25, 31, 34–36, 38, 43, 52, 74, 76–80, 96, 151, 152, 154, 156
, Commissioners (or Board) of 7, 11–14, 17, 43, 143, 144

Longitude (*contd.*),
prize 6, 10, 13
, selenographic 3, 138–140
, terrestrial (or geographic) vii, 2, 8, 16
Lunar distances 6, 11–14, 79, 120

Magnetic variation 8, 16
Mars viii, 51, 149
Mayer's astrolabe 7, 51, 143, 149
, catalogue of zodiacal stars viii, 17, 41, 43, 52, 62, 92, 95, 109, 111, 120, 147, 148, 152, 155, 156
, colour-triangle 45, 52, 122–127, 129, 130, 133, 155, 159
, goniometer 17, 145
, lunar globes 4, 12, 136–138, 142, 160
, lunar map 4, 12, 17, 18, 45, 136–140, 144, 150, 160
, lunar tables 2, 5, 6, 11, 13, 14, 17, 52, 78, 143, 144, 156
, lunar theory viii, 7, 14, 144
, repeating circle 7, 13, 17, 143
, selenographic table 20, 51, 136, 139, 140
, solar tables 2, 14, 17, 42, 94, 143, 156
, theory of the magnet 8, 17, 18, 51, 143
Meridian 3, 28, 30, 41, 42, 68, 69, 72, 74, 78, 92–94, 139, 150, 154, 155
, transit 7, 21, 24, 42, 63, 67, 68, 92–94, 146, 155
Meteorology 53, 58, 113, 114, 158
Micrometer 2, 62, 137
Moon vii, viii, 1–4, 6, 12, 13, 17, 18, 25–28, 34, 36, 38, 72–74, 76–80, 94, 136, 138, 139, 142–145, 153, 154

Nonagesimal 30, 31, 33, 34, 38, 75, 76, 78, 79, 153 154
Nuremberg 2, 3, 5, 10, 12, 134, 136, 142, 149, 160
Nutation 4, 93, 95, 156

Occultation vii, 2, 154
Opera Inedita 12, 17, 19, 43, 45, 46, 145, 149, 150, 161
Osnabrück 16, 17, 20, 145
Observatory of Göttingen, *see* Göttingen Observatory

Parallax 4, 14, 25, 27, 29, 30, 33–38, 71, 74–76, 78, 79, 153, 154
Paris 132, 133, 144, 145, 147, 150, 153, 157, 158
, Academy *Mémoires* 73, 144, 157
Parliament 10, 13
Pendulum clock viii, 3, 7, 16, 67, 68, 92
Precession 4, 44, 93, 95, 156
Prism, glass 81, 82, 154
, wax 124, 125, 134, 135
Projection, geometrical 1, 2, 71–73, 139, 153, 154
Proper motion viii, 18, 24, 43, 44, 52, 96, 109, 110, 114, 148

Quadrant, movable (astronomical) 16, 17, 66, 92, 155
, mural viii, 7, 17, 19, 21, 24, 41, 62–64, 66, 68, 69, 92–95, 120, 121, 143, 148, 152, 153, 155, 159
, nautical 7, 13

Red 81–89, 122–125, 128–132, 154, 155, 159
Refraction, astronomical viii, 5, 14, 21, 33, 92, 95, 119, 121, 149, 156
Right ascension 31, 41, 42, 44, 52, 63, 68, 76, 78, 93–95, 97–108, 110, 111, 156
Royal Academy of Sciences, Göttingen, *see* Göttingen Scientific Society
Royal Society of London 11, 149

St Petersburg Academy of Sciences 5, 143
(Semi-) diameter, of Moon 3, 4, 25, 26, 28, 36, 38, 73, 74, 77–80, 136, 147
, of Sun 38, 39, 73, 74, 77–80
Seven Years' War viii, 7, 10, 11, 15, 149, 155, 156
Shade 40, 81, 84, 88, 91, 135, 139, 155
Solar motion 44, 112, 148
Solstice 41, 58, 60, 94, 151, 152
Stade 16, 18, 145, 149
Star viii, 6, 41, 44, 92–96, 109–112, 114, 147, 148, 154, 156
, zodiacal viii, 6, 7, 17, 62, 92, 95, 120
Strength of colouring 40, 122, 127–130, 147, 159
Sun viii, 3, 38, 41, 42, 44, 54, 58–60, 63, 71–74, 76–80, 93–95, 112, 145, 150–152, 156

Temperature 21, 54–61, 68, 114, 116–118, 150–152
Terrestrial magnetism 8, 143, 149
Thermometer viii, 18, 21, 53–55, 59, 92, 114–116, 118, 119, 121, 148, 157, 158
Thermometric scales 21, 55–59, 61, 114–119, 150, 151, 158

Upsala, *Acta* 73, 154

Venus, transit of 16

White 81, 83, 88, 89, 91, 124, 125, 127, 155

Yellow 40, 81–89, 122–132, 154, 155, 159

Zenith distance 21, 24, 25, 27, 34, 42, 63–69, 93, 94, 120, 121, 159
Zodiac 62, 63, 93, 111